Where Are the Little Green Men?

Terry Adair
Where Are the Little Green Men?

Published by Spines Publishing Platform
ISBN: 979-8-90222-532-4

Spines

Where Are the Little Green Men?

THE ADAIR HYPOTHESIS

TERRY ADAIR

CONTENTS

Prologue: The Rarest Story in the Universe 1
1. The Ingredients of a Habitable World 7
2. The Violent Birth of Stability 17
3. Water: The Great Fortuitous Gift 24
4. The Dawn of Photosynthesis 32
5. Oxygen: The Atmosphere Rewritten 37
6. The Rise of Eukaryotes and the Great Leap Forward 44
7. From Ferns to Forests: The Planet's Carbon Vaults 53
8. Fossil Fuels: Ancient Sunshine in Geological Storage 59
9. The Malthusian Photosynthetic Trap 65
10. Fire: Humanity's First Energy Revolution 71
11. The Limits of Wood, Wind, Water, and Muscle 79
12. Coal: The First Escape from the Trap 86
13. Petroleum: The Liquid Miracle 93
14. Natural Gas: The Cleanest Bridge 103
15. Hydrogen: The Misunderstood Non-Fuel 109
16. Fission: Bottled $E=mc^2$ 116
17. Fusion: Lighting a Star on Earth 134
18. The Great Silence Revisited 142
19. The Adair Hypothesis (Full Exposition) 153
20. The Duty of Civilization 161
Epilogue: The Silent Universe and the Luminous Earth 169

Further Reading & Sources of Inspiration 175
About the Author 177

DEDICATION

For my brother Robin —

whose curiosity and wonder burned bright.

He left too soon to see what humanity may yet become.

"We are the universe made conscious, and life is the means by which the universe understands itself."

— **Brian Cox**

PROLOGUE: THE RARES STORY IN THE UNIVERSE

"A technological civilization can emerge only on a planet that possesses a long-term, accessible surplus of chemically and radiologically stored energy, accumulated through rare evolutionary, geologic, and nucleosynthetic coincidences"

— **Terry Adair**

We live on a small rocky world orbiting an ordinary, middle-aged star. Not at the center of anything. Not in a privileged region of space. Just one planet circling one sun, hanging in one minor arm of a galaxy that looks, from afar, like countless others.

The Milky Way contains roughly 400 billion stars. The observable universe contains two trillion galaxies, each with hundreds of billions more. By any reasonable cosmological accounting, Earth should be forgettable—another speck in an ocean of specks.

Yet this unremarkable world produced something astonishing. It produced us. Creatures capable not simply of survival, but of reflection; not merely of sensation, but of self-awareness; beings who can contemplate quantum theory and galaxies, write symphonies, decipher genomes, split the atom, and send machines to the edge of interstellar space.

For centuries, scientists and philosophers have looked up at the sky and asked a question that feels almost obligatory in a universe this vast: Where is everyone else? Enrico Fermi posed that very question in a lunchroom in 1950, and ever since, countless proposals have been unsatisfactory due to a fundamental oversight—one that will finally be answered.

Given the number of stars, planets, and billions of years available for evolution, it seems natural to assume that intelligent life should be plentiful. Civilizations should be everywhere—some older than ours, some more advanced, some exploring, some communicating, some perhaps even watching.

And yet… Silence. No signals. No visitors. No evidence of ancient empires written in the stars. No signs of the "little green men" that popular culture so casually imagines. Just a quiet cosmos.

This book begins with a simple, yet profound realization:

We may be greatly underestimating how rare a technological civilization truly is.

Not because life itself is rare. Life might be embarrassingly common—microbial mats beneath alien oceans, algae analogues under distant suns, simple organisms thriving in chemical soups scattered across the galaxy.

But civilization—the kind that masters steel, builds engines, transmits knowledge, and shapes its environment—may be one of the rarest phenomena in the universe.

This is the story of why.

It is not enough for a planet to be in the so-called "habitable zone." It is not enough for water to be liquid, for chemistry to be active, or for organisms to emerge. The path from simple life to intelligent civilization is not a single step, but a ladder—with each rung dependent on the one below it, each shaped by a sequence of unlikely cosmic coincidences.

Civilizations do not arise simply from biology. They arise from energy—the ability to capture, store, and deploy it over vast stretches of time. And the universe, for all its apparent abundance, is stingy with the kinds of energy that minds like ours require.

The hinge of this entire story—the fulcrum on which all civilization turns—is something so common on Earth that we rarely pause to acknowledge how miraculous it is: Photosynthesis.

A process that transforms invisible starlight and thin air into solid, structured carbon. The only known mechanism that can trap a star's energy inside matter for millions of years. The quiet alchemy that built forests, oxygenated oceans, sculpted ecosystems, and stocked the Earth with the chemical energy that modern civilization eventually learned to unleash.

Without photosynthesis, there is no biomass. Without biomass, there are no hydrocarbons. Without hydrocarbons, there is no industrial civilization. And without an industrial civilization, there is no science to ask why we are alone.

This book explores that chain of improbabilities. It is a journey from the birth of planets to the birth of minds, from the violence of early solar systems to the quiet mathematics of energy capture, from the chemistry of oceans to the physics of nuclear decay.

It is a story that begins billions of years before humanity existed, in the furnaces of dead stars and the green shimmer of ancient algae, and stretches forward to the present—where a single species on a single world looks up at the sky and wonders how many others ever made it this far.

We may find that the cosmos is full of life… but almost empty of civilizations. And that makes our existence not ordinary, but extraordinary. A rare event in a universe large enough to make "rare" almost indistinguishable from "unique."

This is the story of how—against odds so overwhelming

they defy intuition—we came to be. And why the silence of the cosmos may not be surprising at all.

CHAPTER 1

THE INGREDIENTS OF A HABITABLE WORLD

Stardust and Violence: How to Build a World That Shouldn't Exist

If you step outside on a cold, clear night and look up, you might be tempted to believe the universe is generous. Stars scatter the sky like spilled diamonds. The Milky Way arches overhead in a soft, ghostly river. It all feels so abundant, so overflowing with possibility, that one could almost imagine life springing up everywhere — civilizations scattered across the galaxy like neighborhoods in a cosmic suburb.

But the universe is not generous.

It is selective, stingy, almost unimaginably frugal with the things life requires.

Nearly all the atoms in the cosmos are hydrogen and helium, the two simplest elements, which are believed to be fossil remnants from the Big Bang. Everything else — the carbon in your muscles, the oxygen in your blood, the phosphorus in your DNA, the iron in your hemoglobin, the uranium warming Earth's core — had to be forged in places far more violent than your imagination has ever dared to wander.

Life does not come from starlight.

Life comes from star-death.

Carbon is cooked in the heaving outer shells of dying red giants.

Iron is created in the instant a massive star collapses, when its heart implodes and rebounds as a supernova.

Gold, uranium, and all the heavy elements that underpin civilization are born only in the rarest events the universe offers: the collision of neutron stars, where matter is crushed so violently that even atomic nuclei buckle and fuse.

These events do not happen often.

They seed the galaxy like infrequent punctuation marks in a long, mostly quiet book.

Stars form everywhere — but stars with the right debris around them? Stars whose birth clouds have been seasoned by many generations of cosmic violence?

Those are few.

The Sun was one of the lucky ones.

It formed in a spiral arm of the Milky Way enriched by earlier catastrophes — a neighborhood where ancient giant stars had died, exploded, and strewn their ashes into space. Our solar system began in this graveyard of stellar wreckage, gathering up iron, silicon, carbon, oxygen, phosphorus, sulfur, and the rare radioactive elements that would later keep Earth warm for billions of years.

Had the Sun formed in a calmer corner of the galaxy, far from the ruins of supernovae, it would shine just as brightly — but the planets around it would be barren, starved of the ingredients necessary for continents, oceans, atmospheres, and life.

In other words:

The story of Earth does not begin on Earth.

It begins before the Sun existed.

The Violent Birth of Worlds

Planets do not form gently. The serene orbit you see today — Earth gliding smoothly around the Sun — is the result of a long, brutal adolescence.

Picture a swirling disk of dust and ice, rotating around the newborn Sun. Now imagine that disk turning into a demolition derby. Grains collide and shatter. Pebbles become boulders. Boulders crash into mountain-sized bodies in explosions of molten rock. Entire worlds half-formed are smashed apart and rebuilt.

In this chaos, one collision outshines all the others — the impact between the young Earth and a Mars-sized body we now call Theia. The blow was catastrophic. Much of Earth's crust melted. Thousands of cubic miles of vaporized rock were flung into space, eventually cooling and gathering into the Moon.

It was the most violent event in the history of our planet — and also one of the most fortunate.

For the Moon is not merely a nighttime ornament. It stabilizes Earth's tilt, preventing climate from lurching unpredictably between extremes. Without that stabilizing presence, continents would swing through eras of scorching heat and eternal glaciation. Oceans would slosh unpredictably. Winds would howl in global patterns unlike anything life could easily adapt to.

The Moon is a cosmic shock absorber — and a cosmic stabilizer.

Many planets never experience such a collision.

Many experience too many.

Earth experienced exactly one, and exactly the right one.

The Goldilocks Illusion

So much is made of the "habitable zone" — that neat band around a star where temperatures allow liquid water. It's a comforting concept because it suggests that life should be common. If all you need is to stand in the warm ring around a star, then the universe should be teeming.

But the habitable zone is merely a stage.

It does not write the play.

Being in the habitable zone does not give you:

- an atmosphere
- plate tectonics
- continents
- oceans
- a magnetic field
- long-term internal heat
- a stable rotation
- a Moon
- the right chemical inventory

If the solar system is a neighborhood, Earth did not just get a nice plot of land.

It got the only plot with plumbing, electricity, drainage, and a pantry filled with ingredients no neighboring house received.

Most planets in the habitable zone are sterile deserts, icy wastelands, or steaming pressure cookers wrapped in toxic fumes. Earth's position helped, certainly — but position alone does not explain why we are here.

The Gift of Water — and the Razor's Edge

Water arrived on Earth in pieces — from asteroids, icy bodies, and minerals that released their moisture as the young planet heated. It did not come all at once. It accumulated over hundreds of millions of years.

But here's the miracle:

Earth received the exact right amount.

A little less, and it would be Mars — cold, dry, rusting quietly.

A little more, and it would be a global ocean world, with no continents, no minerals exposed to air, no chance for a carbon cycle, and no stable climate.

Even after Earth received its water, it had to keep it. That required:

- gravity strong enough to hold an atmosphere
- a magnetic field to deflect solar wind
- temperatures cool enough for liquid water
- volcanic outgassing to replenish losses

Mars lost its magnetic field early.

Its atmosphere was stripped.

Its water drifted into space.

Earth retained everything.

Another stroke of cosmic luck.

An Atmosphere Built from Fire

The first atmosphere you would not want to breathe. It was thick with carbon dioxide, nitrogen, water vapor, methane, sulfur dioxide — the fumes of a hot, restless world.

But that atmosphere held heat.

The Sun was dimmer then — only about 70% as bright as today. Without that early greenhouse blanket, Earth would have frozen solid.

CO_2 was not a villain.

It was life's first benefactor.

It kept oceans liquid, weather stable enough, and temperatures within a narrow band that allowed chemistry to begin its delicate dance.

The Invisible Shield

If Earth had no magnetic field, you wouldn't be reading this.

The solar wind — a constant stream of charged particles from the Sun — would have stripped the atmosphere within a few hundred million years. Water would have split into hydrogen and oxygen, and the hydrogen would have bled into space.

Our magnetic field is powered by something that seems almost mystical:

The slow decay of uranium and thorium inside Earth's core.

These rare elements, produced in neutron-star collisions and scattered into the solar system long before Earth formed, supply the heat that keeps the core molten and convective.

That convection drives the dynamo that generates our protective magnetosphere.

Every civilization is ultimately powered by geology — literally.

The Planet That Couldn't Stop Moving

Earth's crust floats on a churning mantle, broken into plates that drift, collide, and recirculate. Plate tectonics is not a universal planetary feature. It might be extraordinarily rare.

Without tectonics, Earth would lose its carbon cycle.

Mountains would not rise.

Oceans would not be replenished with minerals.

Volcanoes would not supply new atmosphere.

The climate would drift into chaos.

Tectonics is the thermostat of life.

It exists because Earth remained warm inside for billions of years.

And that warmth exists because Earth inherited uranium and thorium from dead stars.

Life needs sunlight.

Civilization needs radioactive decay.

The universe provides neither reliably.

Before Life, Everything Had to Work

Only after all these miracles — chemistry, location, water, atmosphere, magnetic shielding, tectonics, internal heat, and stability — could life begin.

Life did not appear in a vacuum of simplicity.

It emerged on a stage constructed by stars, shattered worlds, radioactive elements, gravity, luck, and billions of years of precise circumstance.

By the time the first living cell arrived — perhaps in a vent, perhaps in a tidepool, perhaps in a warm little pond — Earth had passed every cosmic filter.

Most worlds never do.

That is why the first microbial flicker of life on Earth was not merely a biological event.

It was a geological and cosmological triumph.

And from that fragile foothold would come the next miracle

— the invention of photosynthesis — the moment life learned how to steal the power of the Sun and store it in matter.

A moment that would change everything, forever.

But that belongs to Chapter 4.

CHAPTER 2

THE VIOLENT BIRTH OF STABILITY

The transition from chaos to calm is never a clean break. It is a negotiation—violent, protracted, and costly. In the early history of Earth, stability did not arrive gently, nor did it resemble the serene, temperate world we know today. What emerged was something subtler: a planet that finally granted time. Time for oceans to stop boiling away. Time for chemistry to endure more than a fleeting moment. Time—after millions of years of cosmic assault—for possibility to take root.

This chapter tells the story of how Earth won that time.

The giant impact that reshaped our planet, the formation of the Moon, the gift of seasons, the birth of a magnetic shield, the slow cooling of a radioactive interior—each was an act of violence that paradoxically carved out the conditions for eventual tranquility. The

Earth that emerged from these catastrophes was not simply habitable. It was maintainably habitable. It possessed not only liquid oceans but also the long-term stability required for life to emerge, persist, and ultimately flourish.

This is the story of that stability's unlikely birth.

Theia: The Catastrophe That Saved the Planet

Roughly 4.5 billion years ago, an object about the size of Mars—now called Theia—collided with the young Earth. "Collided" hardly captures the scale. This was not an impact; it was a planetary merging event, a cosmic surgical strike that ripped away Earth's outer layers and vaporized both bodies in a single incandescent moment. What fell back together was not the old Earth, but a new one—larger, hotter, and chemically altered.

The debris that escaped Earth's gravity coalesced into the Moon, our companion whose influence has been nothing less than foundational:

- It stabilized Earth's axial tilt, preventing the planet from toppling chaotically like Mars.
- It created the tides, stirring early oceans and accelerating chemical exchange.
- It slowed Earth's rotation, softening violent atmospheric winds.
- It forever locked Earth into a predictable day–night rhythm.

Without Theia, Earth might have remained a geological delinquent—spinning wildly, shearing through climate cycles, and denying chemistry any hope of continuity. Instead, catastrophe gave rise to order.

Seasons From Violence

Earth's axial tilt—approximately 23.5 degrees—was a direct consequence of the Theia impact. That tilt gave us seasons, which in turn prevented the planet from falling into climatic stagnation. Seasons are not decorative; they regulate the cycling of heat, water, and nutrients. They create environmental diversity.

A planet without seasons risks becoming a world of extremes: frozen poles locked permanently in night, equatorial belts scorched by perpetual sunlight, and a biosphere constrained to narrow margins. However, Earth's tilt ensured dynamic circulation patterns that stirred the oceans, sculpted the winds, and distributed energy across latitudes.

In the aftermath of violence, Earth inadvertently acquired a thermostat.

A Magnetic Shield Forged in Fire

Violence also bequeathed another essential gift: Earth's magnetic field. When the young Earth differentiated into core, mantle, and crust, the iron-rich core remained molten—heated from within by leftover accretion energy and sustained by long-lived radioactive decay.

The churning of this molten metal generated a magnetic

dynamo that shielded our atmosphere from the solar wind. Without it, the Sun would have stripped Earth's air away molecule by molecule, as it did to Mars.

The magnetic field was not a luxury; it was a survival mechanism. It protected not only the atmosphere but also the early chemistry experimenting within it—preventing the erasure of every fragile chemical advance.

Radioactive Heat: The Planet's Internal Engine

Buried within the Earth's interior were long-lived isotopes—uranium-238, uranium-235, thorium-232, potassium-40. Their decay did more than generate heat. It maintained a convective mantle, a molten core, and a restless crust.

This was not incidental. Without internal heat, Earth's surface would have become geologically dead long before life ever emerged—cold, stable, and chemically barren.

Instead, the radioactive embers kept the planet alive from the inside, driving volcanism, powering hydrothermal systems, and creating an endless supply of fresh crust and mineral diversity.

Plate Tectonics: The Slow Choreography of Habitability

With a hot interior and cooling surface, Earth developed a global tectonic engine—a continuously shifting puzzle of rigid plates floating atop a ductile mantle. Plate tectonics would become one of the most important planetary processes in life's long-term survival:

- It recycled carbon through subduction and volcanism.
- It created continents and shallow seas rich in minerals.
- It stabilized global temperatures through the carbonate–silicate cycle.
- It maintained a dynamic interface between the ocean and crust, where chemistry could be experimented with.

A planet without plate tectonics is a planet without long-term climate control. Earth's cycles—ice ages, warm periods, the slow pulse of atmospheric carbon—were not happenstance. They were engineered by geology.

The violent forging of Earth's interior did not simply produce mountains. It produced stability.

A World No Longer Trying to Kill Chemistry

After hundreds of millions of years of bombardment, a quieting fell. Not peace, but predictability.

For the first time:

- Oceans remained oceans instead of evaporating into steam.
- Atmospheres endured long enough for molecules to interact with one another.
- Minerals accumulated faster than they were obliterated.

- Thermal cycles repeated instead of resetting with each collision.

Life did not require calm—it required consistency. A universe may create stars in chaos, but biology requires order and rhythm. The early Earth finally began to offer them a home.

This moment—this narrow window when Earth stopped resetting its own progress—marks the true beginning of habitability. Not the presence of water, nor the right orbit, nor the temperate surface. But the gift of time.

Life's First Environment: Not Sunlight, but Stone

Popular imagination places the origin of life in sunlit pools or warm ponds. Reality points somewhere else—deeper, darker, more primeval.

Life's first crucible was likely in places where Earth's internal fire met its cooling oceans:

- alkaline hydrothermal vents
- mineral-rich fissures in newly formed crust
- sulfide chimneys acting as natural electrochemical reactors
- porous stone labyrinths where gradients of heat and chemistry created persistent cycles

These environments offered something the surface could not: stability. Even as meteorites still occasionally fell and lightning split the skies, the deep ocean offered shelter. More importantly, it offered continuous energy. Chemical energy derived from rock-water interactions.

Life's first spark was not sunlight but geochemistry. Before photosynthesis, before membranes, before nucleic acids refined their copying tricks, Earth provided tiny mineral-bound laboratories where matter could practice persistence.

The Slow Rise of the First Microbial World

Life did not appear fully formed. It emerged as a sequence of solutions—chemical, structural, and energetic. Early molecules learned to last just a little longer. Some arrangements were copied imperfectly. Some managed to enclose themselves, not with modern membranes, but with mineral scaffolding or simple hydrophobic films.

The first microbial world was not a kingdom of cells, but a whisper of chemistry that learned how to avoid dissolution.

Persistence became competition. Competition became refinement. And refinement, over millions of years, evolved into autonomy.

Earth's violent beginnings had finally given way to the quiet, patient work of chemistry becoming biology.

CHAPTER 3

WATER: THE GREAT FORTUITOUS GIFT

Earth's first water did not fall from the sky. It was born in darkness, locked inside minerals before the planet even existed as a coherent world. Long before oceans shimmered blue under sunlight, hydrogen and oxygen were trapped inside dust grains forged in ancient stars. These grains drifted through the protoplanetary disk, coating themselves with thin films of ice in the cold outskirts, only to be carried inward toward the newborn Sun, where they were shattered, melted, and folded into the interior of the forming Earth.

The earliest Earth — the planetesimal that would one day soften into continents and seas — contained water in the form of hydroxyl groups bonded to silicate minerals. Olivine, pyroxene, and even early basaltic melts carried hydrogen deep beneath the surface. In these minerals, water was not a liquid but a chemical memory: atoms held in crystal lattices, waiting for a world cool enough to let them breathe.

As Earth grew, accreting mass through one collision after another, much of its inherited water evaporated into space. The heat of formation was so intense that any free molecules of H_2O were vaporized. Yet the mineral-bound water endured. Even in a magma ocean, rocks held on to a fraction of their hydrogen. And when Earth finally differentiated into core, mantle, and crust, these water-bearing minerals were dragged downward, dissolving into molten rock that still contained the signature of ancient ices.

Earth's mantle today still holds several oceans' worth of water — not as liquid seas but dissolved, atom by atom, in minerals born under crushing pressure. Deep inside, water is not something the Earth received. It is something Earth was.

But this interior water was only part of the story. The Earth's original endowment was insufficient to fill the oceans, which are thousands of kilometers wide. Much of the water we now know — the water of rivers and lakes and storms and tears and blood — arrived later, delivered from the outer solar system by a rain of icy bodies that fell with increasing intensity during a singular epoch of destruction and renewal known as the Late Heavy Bombardment.

Roughly 4.1 to 3.8 billion years ago, the solar system was reshaped in a gravitational convulsion. The giant planets, particularly Jupiter and Saturn, migrated through resonances that destabilized vast populations of icy planetesimals in the Kuiper Belt and the remains of the primordial asteroid belt. These bodies, composed of silicates, organics, carbon dioxide, ammonia, and most critically, water ice, were hurled inward toward the terrestrial planets like shrapnel flung from a cosmic explosion.

During these turbulent eons, Earth was struck by objects containing enormous stores of water — carbonaceous chondrites, hydrated asteroids, outer belt bodies rich in phyllosilicates, and occasional comets whose crystalline ices carried isotopic signatures born in the colder reaches of the disk. Every impact was catastrophic, releasing energies that dwarfed nuclear weapons by magnitudes. Oceans boiled. Crusts melted. Atmospheres were blasted away. But as the planet radiated heat back into space after each cataclysm, new water condensed again from the steam, adding to the reservoirs that slowly deepened across its cooling surface.

The retention of this water was no trivial matter. Earth was large enough, massive enough, and cool enough to hold onto vapor without letting it escape to the vacuum. Hydrogen, the lightest element in the universe, ordinarily flees into space unless trapped by gravity or shielded by an atmosphere. But Earth's gravity, enhanced after the Theia impact, was strong enough to prevent wholesale loss, and the thickening atmosphere slowed photochemical escape. The presence of a magnetic field, generated by a convecting iron core, protected Earth's skies from the solar wind that stripped Mars. In that magnetic cocoon, Earth's atmosphere endured long enough for water to settle, accumulate, and eventually shape the very face of the planet.

Even as these oceans formed, they were far from stable. The newborn Sun was faint — roughly 70% of its current luminosity — and without greenhouse gases, Earth would have been a frozen orb. But early volcanic outgassing produced a dense atmosphere of carbon dioxide, methane, nitrogen, and water vapor, providing the radiative insulation necessary to keep the oceans liquid. The interplay between greenhouse gases and ocean heat capacity, between volcanic release and carbon burial, between mantle hydration and surface condensation, produced a delicate thermal balance. It was not guaranteed. It was earned, step by precarious step.

To appreciate the fragility of this balance, one must look to Earth's siblings.

Mars, too small and too cold, lost its magnetic dynamo early. Without a shield, the solar wind stripped the atmosphere down to a whisper. Even if Mars once held rivers, lakes, and shallow seas — and evidence suggests it did — the loss of atmospheric pressure doomed its water to escape or freeze. The planet dried from the outside inward, leaving only scattered remnants of ice locked in soil and caps.

Venus, nearly Earth's twin in size, met the opposite fate. Closer to the Sun, its surface warmed until its early oceans evaporated. Water vapor — a powerful greenhouse gas — trapped more heat, which accelerated evaporation in a devastating positive feedback. Ultraviolet radiation split the water in the upper atmosphere; hydrogen is lost to space, and oxygen is absorbed into rocks. Venus did not simply lose its oceans. It erased the very conditions that would have allowed oceans to exist again.

Earth alone walked the fine line. It had enough gravity to hold an atmosphere. Enough volcanism to sustain greenhouse warming during the faint young Sun. Enough distance to prevent runaway evaporation. Enough internal heat to maintain a magnetic shield. Enough mantle hydration to sustain long-term volcanism and plate tectonics. Enough surface temperature to allow water to condense into stable seas.

And so, slowly, Earth transformed from a molten world into a blue one. Oceans pooled in basins, covering most of the surface. Rain carved channels into young continents. Hydrothermal vents sprayed mineral-rich water into deepening seas. Tides, guided by the Moon, churned the coasts. Rivers delivered dissolved ions from granite and basalt, feeding the chemistry of early life. Storms, waves, and currents became global engines of mixing — a planetary mixer in which molecules could meet, react, assemble, and persist.

Water did something profound:

it gave Earth time.

Time for the early atmosphere to stabilize.

Time for minerals to cycle.

Time for continents to form.

Time for chemistry to turn into biology.

Mars ran out of time.

Venus heated past it.

We inherited it.

The water of Earth is an archive of planetary history. Some were trapped in the rocks from which the planet formed. Some arrived frozen, at the tail end of the solar system's most violent era. Some was outgassed from the mantle as volcanoes roared through eons. And some is still cycled, dissolved, subducted, squeezed, and returned in a dance between mantle and ocean that has continued unbroken for billions of years.

Earth's oceans were not gifted once. They were maintained — against fire, against freezing, against gravity, against escape, against time.

And because they endured, life endured.

Because they persisted, complexity persisted.

Because they remained liquid under a changing Sun for billions of years, evolution had a stable canvas on which to paint.

Water on Earth was not inevitable.

It was improbable.

Fragile.

Fortuitous.

It arrived in ice and steam, in mineral lattices and ancient impacts. It survived cataclysms that erased continents. It outlasted the Sun's dim youth and its brightening maturity. It threaded the needle between Mars and Venus — too little and too much, too cold and too hot, too weak and too fierce.

Earth kept its water long enough for life to rise, for life to change the planet, and for the planet to give birth to minds capable of understanding what an astonishing miracle water truly is.

In the stillness of Earth's early oceans, water carried not only heat and minerals but possibility. Every current, every droplet, every hydrogen bond formed a stage upon which chemistry could explore its limits. And though life's earliest metabolisms drew their energy from the rocks and vents below, the greatest energy source of all still shone above them, untouched. The Sun watched over a world not yet ready to use its power. But life, persistent and curious, would eventually find a way to turn starlight into sustenance—and in doing so, would change the destiny of the planet.

Addendum: Scientific Insight: Bennu and the Common Chemistry of Life

NASA's OSIRIS-REx mission returned pristine material from the carbon-rich asteroid Bennu — a small, randomly sampled remnant of the early Solar System. Laboratory analyses revealed a remarkable inventory of prebiotic compounds, including:

- All five nucleobases used in DNA and RNA
- Multiple protein-forming amino acids, 14 of 20
- Hydrated minerals and carbonate salts indicate ancient liquid water
- Presolar grains older than the Sun itself

Bennu's chemistry shows that the building blocks of life arise naturally in small, water-containing bodies formed during planetary assembly. Far from being rare, these organics appear to be widespread byproducts of Solar System evolution. If a half-kilometer rubble pile contains such molecular richness, countless similar objects likely do as well.

This strengthens a central conclusion of modern cosmochemistry: life's foundational molecules are ubiquitous, not exceptional. The early Earth did not acquire its prebiotic inventory through a singular stroke of luck, but through participation in a larger cosmic pattern — one in which water, carbon, and energy routinely collaborate to assemble the alphabet of biology.

CHAPTER 4

THE DAWN OF PHOTOSYNTHESIS

How Early Microbial Life Transformed Sunlight Into Chemical Power

Before chlorophyll, before pigments danced with photons, before life learned to touch the Sun, Earth's biosphere was a dim, geochemical struggle. Microbial communities scraped energy from minerals and vents, surviving on the chemistry of stone and heat. The oceans were dark not only to the eye—there were no eyes—but dark in spirit, a world running on the faint spark of planetary metabolism.

And then, over hundreds of millions of years, life invented a way to tap a power source that dwarfed anything the young Earth could offer.

It began modestly. Early bacterial pigments—likely precursors to chlorophyll—absorbed stray photons not to make food, but to shield fragile molecular structures from ultraviolet radiation. In a world without ozone, sunlight was a hazard, not an opportunity. These pigments were accidental guardians.

But evolution is the great thief of accidents. Molecules that absorbed light could, under the right structural quirks, funnel that energy into reactions instead of merely dissipating it. And so began a long, improbable transition: light absorption coupled to electron transfer, electron transfer coupled to energy storage. The biosphere began learning the grammar of light.

The emergence of chlorophyll—with its elegant porphyrin ring and magnesium heart—was the moment when that grammar became poetry. Chlorophyll is not merely a molecule; it is a bridge, reaching from the furnace of a star to the machinery of a cell. When a photon strikes it, an electron is lifted into a higher energy state. That electron, racing away from its previous equilibrium, can be captured by protein complexes that perform the delicate art of charge separation. In that instant, sunlight becomes chemistry.

The earliest photosynthetic organisms were anoxygenic bacteria, using molecules like hydrogen sulfide rather than water as electron donors. These microbial lineages—purple bacteria, green sulfur bacteria, heliobacteria—represent ancient, primordial branches of the photosynthetic tree. Their world was one where light was sparse and chemical donors were precious. Yet these bacteria flourished, weaving themselves into sediment mats, increasing the biosphere's reach, widening the base of life's pyramid.

But the greatest biochemical invention in Earth's history—greater than the rise of complex cells, greater than multicellularity, greater even than nervous systems—was still to come.

Somewhere between 3.4 and 2.7 billion years ago, a particular lineage of bacteria—ancestors of modern cyanobacteria—assembled a molecular machine capable of performing one of the most demanding reactions in nature: the splitting of water.

Photosystem II appears to have emerged only once. The evidence is overwhelming. Its structure, complexity, and catalytic core—the manganese-calcium-oxygen cluster that tears electrons from the bonds of water—betray a long co-evolution of many parts. It is so improbable that no evolutionary pathway has ever duplicated it. Flight arose at least four times: bats, birds, pterodactyls and insects. Vision evolved independently in dozens of lineages. Echolocation appeared in both bats and whales. But oxygenic photosynthesis stands alone: a singular, unrepeatable stroke of biochemical genius.

And its appearance changed everything.

Water is the most abundant molecule in the oceans. To find a way to strip electrons from it was to unlock an energy source effectively without limit. With Photosystem II, cyanobacteria gained access to a river of electrons as vast as the seas themselves, powered by the blazing generosity of the Sun.

The byproduct of this miracle was a highly reactive gas that had been scarce on Earth since its formation: oxygen.

In the earliest days of oxygenic photosynthesis, this oxygen was not free. It reacted instantly with iron dissolved in the oceans, forming rust that fell to the seafloor in bands of red and black sediment—the silent geological footprints of a metabolic revolution. The oceans drank the new gas greedily, and the atmosphere remained unchanged for millions of years. The world was not yet ready for oxygen. That chapter lies ahead.

For now, what matters is this: photosynthesis transformed sunlight into a stable, storable reservoir of chemical energy. Every molecule of carbohydrate is a ledger of photons, a diary of the Sun inscribed in carbon bonds. Every cell built by photosynthesis carries the memory of starlight captured by chlorophyll and braided into matter. All forests, all fields, all crops, all ecosystems rest upon this single biochemical triumph: biomass as stored starlight.

With chlorophyll's invention and the rise of Photosystem II, Earth stepped from geochemical austerity into energetic abundance. Life's power source was no longer the deep crust —it was the sky.

Sunlight, once a hazard, had become the planet's greatest fortune.

The full consequences of that audacity—the rewriting of the atmosphere, the reshaping of oceans and minerals, the rise of aerobic respiration, and the path toward complexity —belong to the next chapter. But the seed of that transformation was sown here.

In the dawn of photosynthesis, life reached for the Sun and found the power to change a world.

CHAPTER 5

OXYGEN: THE ATMOSPHERE REWRITTEN

How oxygen reshaped oceans, minerals, metabolism, and the evolution of complexity. The combustion paradox: oxygen as both creator and destroyer.

Oxygen began as a rumor—an unassuming byproduct of microbial ambition, released in quantities so small that the early Earth swallowed it whole. For hundreds of millions of years, every molecule of O_2 produced by cyanobacteria vanished instantly into the planet's chemistry. The newborn gas had no future, no accumulation, no destiny. It reacted with dissolved iron, transforming the oceans from ink-dark to blood-red as layers of rust drifted to the seafloor. It oxidized volcanic gases. It sank into sediments. It disappeared.

But oxygen never stopped being made.

Every sunrise, the cyanobacteria wove more starlight into

sugar, more electrons into new metabolic pathways, more oxygen into oceans that struggled to absorb the rising tide. Photosystem II continued its quiet alchemy, splitting water, hurling electrons into biochemical cascades, and discarding oxygen as an unwanted side effect. What began as a molecular accident became a planetary negotiation.

And eventually, the planet lost the argument.

When the sinks filled—when the iron was exhausted, when the oceans reached saturation, when the crust had drunk all it could—oxygen crossed a threshold. A tipping point. A geochemical rebellion. One day, sometime around 2.4 billion years ago, there were no more places for the gas to go.

And on that day, oxygen entered the sky.

Its ascent was not a sudden flash but a slow ignition. Fraction by fraction, molecule by molecule, oxygen began to accumulate in the atmosphere. The air changed, and with it, the destiny of life. Earth, once a world of methane, sulfur, and carbon dioxide, began its transformation into a blue planet.

This is known to geologists as the Great Oxygenation Event, but its true name is simpler: the atmosphere rewritten.

The consequences were immediate and profound. Methane, once abundant and warming, reacted with oxygen to form carbon dioxide and water. The greenhouse blanket thinned. Temperatures fell. Some models suggest that the planet slipped into its first global glaciation—a world encased in ice —triggered by the metabolic success of microbes no thicker than a human hair.

In the oceans, ancient anaerobic kingdoms faced a slow, suffocating doom. Oxygen is a ruthless molecule. It tears apart chemical bonds, cripples pathways that evolved in its absence, and renders entire lineages chemically obsolete. To the anaerobic world, oxygen was poison—an atmospheric apocalypse.

Yet even as it destroyed, oxygen created.

The paradox of oxygen is that it is both a bringer of death and a giver of extraordinary life. In its reactivity lies its danger—but also its gift. Cells that learned to tame oxygen gained access to a furnace of biochemical power. Aerobic respiration, which uses oxygen to extract energy from organic molecules, is thirty times more efficient than the anaerobic pathways that preceded it. A single glucose molecule fed into oxygen-driven metabolism releases a flood of ATP—a wealth of chemical energy that fuels motion, growth, and eventually complexity.

Without oxygen, no organism could afford the metabolic cost of a large body, an active lifestyle, or a hungry brain. Oxygen is the reason complexity has the freedom to exist.

Geologically, oxygen created new minerals. Before its rise, Earth's mineral diversity was modest. But oxygen reacts with almost everything. It oxidizes metals into dazzling colors, carves gemstones into brilliance, and expands mineralogical possibilities into the thousands. Ruby, sapphire, malachite, turquoise, hematite—these are gifts of an oxygenated world.

It also built the sky's shield. As oxygen accumulated, ultraviolet radiation transformed a fraction of it into ozone, forming the thin atmospheric veil that allowed life to venture from the oceans without being sterilized by the Sun. Without ozone, continents would forever remain barren. With it, Earth became a canvas awaiting the brushstrokes of land-based life.

However, the most profound consequence of oxygen was one that occurred within living cells. Oxygen forced evolution to confront a biochemical dilemma: how to contain a molecule powerful enough to destroy the very structures that harnessed it. Oxygen is the parent of fire, and metabolism is controlled combustion. Life learned to handle this fire with delicate precision, binding oxygen to enzymes, shielding vulnerable molecules from its reactive claws, and channeling its fury into useful work.

Every aerobic cell is, in a sense, a tiny forge. Within membranes, electrons flow like sparks, cascading down gradients toward oxygen, releasing energy in controlled bursts. Life became a dance of electrons seeking their final, oxygen-filled home.

And from this metabolic revolution came the possibility of large, complex cells—the eukaryotes who would later build tissues, organs, bodies, and eventually civilizations. But that is a later chapter. For now, oxygen set the stage. It rewrote the rules. It shifted the energy economy of Earth from subsistence to abundance.

Yet the paradox remains: oxygen is life's most generous ally and its most dangerous enemy. It fuels the fire of metabolism, but it also corrodes the molecules of life. It builds, and it erases. Every organism that uses oxygen lives on the edge of a chemical cliff, protected by a delicate balance of antioxidants, enzymes, and membranes that tame the reactive flame.

For all its promise, the rise of oxygen was not a gentle dawn. It was a catastrophe. The first great die-off in Earth's history unfolded quietly and without witnesses, yet its magnitude rivals the worst extinctions life would ever experience. Entire kingdoms of anaerobic organisms—microbes that had thrived for over a billion years—found themselves suffocated in their own world.

Oxygen seeped into environments that had never known it. To the anaerobes, it was not a gift but a toxin—an invisible, relentless assassin. Their enzymes unraveled in its presence. Their metabolic pathways collapsed. Their cell membranes oxidized and dissolved. The ancient biochemistry that had once powered the planet's first ecosystems became chemically obsolete in a world turning blue.

One by one, microbial strongholds fell. Mats that had carpeted the seafloor vanished. Vast populations dwindled to refuges in oxygen-free sediments, brines, and deep subsurface pockets where the new air could not reach. The old world died so the new world could breathe.

This was not a cleansing—it was a planetary redefinition. The oxygenation of Earth did not merely add something new; it removed nearly everything old. It was a transition as violent as any asteroid strike, but its weapon was a molecule. Its force was slow, implacable, and total.

In the end, only two kinds of life endured: those few anaerobes fortunate enough to find sanctuary,and the innovators —those rare lineages that learned not only to survive oxygen, but to wield it as a source of extraordinary power.

From this biochemical purge arose the potential for everything that would follow. Complexity. Respiration. Movement. Predation. Brains. All of it was born from a die-off so profound that it reset the evolutionary stage and wrote the rules of life anew.

The Great Oxygenation Event was not just the atmosphere rewritten; it was a rewriting of the atmosphere.

It was the biosphere reborn.

The atmosphere we breathe is not merely air. It is the residue of ancient microbial labor, the breath of cyanobacteria written across the sky. It is the inheritance of a world transformed by a molecule that almost killed everything before, yet powered everything that came after.

The Great Oxygenation Event was not just a geological transition. It was the moment Earth became a world where complexity was possible, where sunlight could be stored in forests, where animals could one day roam, where minds could one day think.

Oxygen opened the door.

Life would spend the next two billion years learning how to walk through it.

CHAPTER 6

THE RISE OF EUKARYOTES AND THE GREAT LEAP FORWARD

Endosymbiosis, mitochondria, chloroplasts, and the emergence of multicellular life. Biology harnessing energy at higher efficiency.

For billions of years, life existed in a narrow biochemical corridor. The only organisms on Earth were prokaryotes—simple cells, efficient and resilient, but locked behind an energetic wall they could not breach. They ruled the planet's oceans, sediments, and hydrothermal seams, yet in all their microbial glory, they remained small, solitary, and constrained by the laws of thermodynamics.

No matter how clever their metabolisms, no matter how diverse their enzymatic tricks, prokaryotes lived in a world defined by limits. A bacterial cell can grow only so large before diffusion slows to a crawl. It can store only so much energy before its membrane potential falters. It can sustain only so many internal reactions before crowding becomes fatal. It operates on a perpetual austerity budget—nimble, yet energetically impoverished. For nearly two billion years, this ceiling held firm.

There were no bodies.

No tissues.

No nervous systems.

No neurons or muscles or multicellular architectures shaping the water column.

Had evolution ended here, Earth would have remained a microbial world indefinitely—an ocean planet shimmering with invisible life but devoid of forests, animals, and the great carbon burial events that would one day make civilization possible. The transition out of this prokaryotic trap required an event that defied the typical patterns of evolution: not a gradual accumulation of traits, but a merger—a biological unification so improbable that it seems less like invention and more like a cosmic accident.

Around two billion years ago, an archaeal cell—a predator among microbes, capable of engulfing other cells—swallowed a bacterium. This encounter should have been lethal. One cell consuming another was a prelude to digestion. But evolution does not follow scripts. The swallowed bacterium did not die. It persisted inside its host, continued its metabolism, and eventually became indispensable.

Over time, this internalized bacterium evolved into the

mitochondrion—the biochemical engine that enables complex life. The merger was nothing short of revolutionary. Mitochondria take organic molecules and oxygen and drive electrons through a chain so efficient that its output dwarfs every other known metabolic strategy. Where anaerobic pathways yield meager energy, mitochondria unleash torrents of ATP—orders of magnitude more than a prokaryote could ever hope to produce.

With this surplus, the host cell could now build internal membranes, regulate its internal chemistry with unprecedented precision, and support far more genetic information. It could grow larger without collapsing under the constraints of diffusion. It could develop a cytoskeleton to change shape, engulf prey, and explore new ecological roles. Complexity—not as metaphor, but as function—became thermodynamically affordable.

This was the moment the energy ceiling shattered.

What followed was not a rapid bloom, but a long, deliberate rise. Eukaryotes experimented with new architectures, including flexible membranes, dynamic internal scaffolding, and intricate gene regulation. They diversified into predators, grazers, and symbiotic partners. The planet was still largely microbial, but a new kind of cell was taking shape—a cell with compartments, power plants, and the ability to build larger bodies and more intricate systems.

Hundreds of millions of years after the mitochondrial merger, a second endosymbiotic event occurred—equally improbable, equally transformative. A eukaryotic cell engulfed a cyanobacterium, one of the ancient masters of photosynthesis. Once again, digestion failed to occur. The captive cyanobacterium took up residence inside its host, and over time, the two fused into a new entity: the chloroplast.

The chloroplast retained echoes of its past—its own genome, its own membranes, its own evolutionary lineage—but it was now part of a larger cell, one capable of building tissues, colonies, and ultimately forests. With chloroplasts, eukaryotes could tap sunlight with unprecedented efficiency. This was photosynthesis scaled up—no longer confined to the free-floating mats of microbes but embedded in the architecture of growing, dividing, increasingly complex organisms.

The oceans transformed as algae proliferated. Food webs lengthened. New ecological roles emerged. For the first time, photosynthesis and cellular complexity coevolved within the same lineage, creating organisms capable of capturing carbon at a planetary scale.

This pairing—mitochondria providing energy density, chloroplasts providing energy capture—set the stage for one of the most consequential transformations in Earth's history: the rise of multicellular life. Large bodies demand reliable energy sources. Tissues demand coordination. Development demands an internal economy far more sophisticated than any prokaryote could sustain. With mitochondria and chloroplasts, cells could afford these costs.

Evolution began to build on this new foundation. Simple colonies formed, then sheets of cells, followed by bodies with distinct fronts, backs, symmetry, and specialization. A billion years ago, multicellular algae were constructing the first marine landscapes. By the time the earliest animals appeared, Earth had already become a planet of metabolically empowered cells—cells able to grow, differentiate, and cooperate in ways that reshaped every ecosystem.

The rise of eukaryotes was not just a biological revolution. It was an energetic revolution. Before mitochondria, life lived paycheck to paycheck. After mitochondria, life had savings, investments, and the surplus needed for innovation. Before chloroplasts, photosynthesis was a thin microbial veneer. After chloroplasts, photosynthesis became a planetary force.

And from these intertwined lineages came everything that would matter for the carbonization of Earth: forests, plankton blooms, food chains, sedimentation, burial, petroleum precursors, coal swamps, and the long-term sequestration of carbon that would one day fuel the fires of industrial civilization.

Endosymbiosis happened twice. Only twice. In four billion years of evolution, the universe offered this gift to life just two times on one world orbiting one star. Without these mergers, there is no multicellularity, no biomass surge, no carbon burial, no hydrocarbons, no metallurgy, no physics, no escape from the microbial ceiling.

The rise of eukaryotes was the great leap forward—a leap across an energetic chasm that trapped most worlds forever in microbial stillness. On Earth, this leap opened the path to everything that followed.

It was the moment complexity became possible.

It was the moment Earth began preparing the fuel for civilizations yet to come.

Hundreds of millions of years after the mitochondrial merger, evolution repeated its most unlikely magic. A eukaryotic cell engulfed a free-living photosynthetic bacterium. But this was not just any microbe—this was a cyanobacterium, one of Earth's most ancient and transformative organisms. Cyanobacteria were the architects of oxygenic photosynthesis, the microbes that had rewritten the atmosphere and rusted the oceans red.

What happened next was not symbiosis in the casual sense; it was the permanent fusion of two lineages. The engulfed cyanobacterium took up residence inside its host and, over long stretches of evolutionary time, surrendered much of its autonomy. Genes migrated into the host's nucleus. Metabolic pathways intertwined. Membranes reorganized. What remained was no longer a bacterium living inside another cell—it was a chloroplast, the photosynthetic organelle found in every plant, every alga, every leaf that has ever unfurled under the Sun.

A chloroplast is not "like" a cyanobacterium.

It is a cyanobacterium—pared down, specialized, integrated, but still bearing its own circular genome, its own bacterial ribosomes, its own memories of a 3-billion-year evolutionary past. Under the microscope, its ancestry is unmistakable. Every blade of grass is, in a biochemical sense, a forest of domesticated cyanobacteria. Every giant redwood is a towering alliance between eukaryotic architecture and bacterial photochemistry. Photosynthesis in modern plants is not a new invention; it is the ancient metabolism of cyanobacteria scaled up built within bodies that evolved long after the merger.

This second endosymbiosis—this domestication of a free-living cyanobacterium—was one of the most consequential events in Earth's history. It allowed eukaryotes not merely to survive, but to harness sunlight on a planetary scale, setting the stage for forests, algal blooms, food webs, and the vast sequestration of carbon that would one day shape the climate, geology, and civilization itself.

Where mitochondria shattered the energy ceiling, chloroplasts illuminated the stage on which life would finally grow tall.

For more than half a century, biology recognized only two clear, transformative cases of primary endosymbiosis: mitochondria and chloroplasts — ancient free-living bacteria captured and permanently integrated into eukaryotic cells.

Until recently, these events appeared unique.

However, compelling evidence now supports a third, exceedingly rare example: the *nitroplast.*

Nitroplasts originate from a nitrogen-fixing cyanobacterium (*UCYN-A*) that entered into a stable endosymbiotic relationship with a marine alga (*Braarudosphaera bigelowii*). Over evolutionary time, the symbiont lost independence: it shed large portions of its genome, transferred genes to the host nucleus, synchronized its division with the host cell cycle, and became metabolically indispensable.

Most notably, the nitroplast performs a function previously unknown in eukaryotes — biological nitrogen fixation — converting atmospheric nitrogen into bioavailable forms directly inside the host cell.

Crucially, this does not overturn the broader conclusion of this chapter.

Endosymbiosis remains extraordinarily rare.

It has occurred only a handful of times across billions of years of life.

Each instance required precise metabolic compatibility, long-term stability, and evolutionary patience on geological timescales.

Rather than weakening the endosymbiotic theory, nitroplasts strengthen it — demonstrating that while such integrations *can* occur, they do so only under exceptionally constrained conditions.

Endosymbiosis is not a frequent innovation.

It is a biological anomaly.

And when it happens, it reshapes the trajectory of life itself.

CHAPTER 7

FROM FERNS TO FORESTS: THE PLANET'S CARBON VAULTS

The few hundred million years in which Earth buried the energy of ages, building the carbon reservoirs that would fuel continents, climates, and civilizations.

Life had conquered the oceans long before it ever touched the land. For billions of years, Earth's continents stood silent beneath a sky shaped by microbes, barren stone washed by tides, unmarked by roots or leaves or the whisper of wind through branches. The landmasses of the early Paleozoic were chemically rich but biologically void—a world of potential waiting for the right biochemical tools, the right biological courage, the right evolutionary gamble.

That gamble began around 470 million years ago, when a lineage of green algae made the slow, improbable transition from shallow waters onto damp shorelines. These pioneers were tiny, fragile things—threadlike strands and moss-like mats that clung to moisture and feared the drying power of sunlight. But they brought with them something unprecedented: the chloroplasts of cyanobacterial ancestry, the machinery of photosynthesis refined over eons in the sea. The moment plants stepped onto land, Earth gained, for the first time in its history, photosynthesis operating on a continental scale.

The initial colonizers were not forests. They were carpets—low mats of liverworts and early bryophytes that hugged the ground. But even these modest forms began to alter the continents. Their rhizoids bound sediments. Their metabolic processes initiated the first stages of soil formation. Their photosynthesis drew carbon dioxide out of the air and into organic matter that resisted decay. And their presence invited the next evolutionary steps.

As life continued to push deeper onto land, evolution began to construct new architectures. Around 425 million years ago, the first vascular plants appeared. These were no longer simple mats; they were upright, internal-plumbing-bearing organisms, strengthened by lignin—one of the most complex molecules life has ever produced. Lignin made stems rigid, resistant, and capable of lifting them, allowing the plants to lift their leaves into the air where sunlight was more abundant. The invention of lignin marked a biochemical revolution, enabling plants to grow taller.

Roots followed, delving into sediments, fracturing rock, and accelerating weathering on a global scale. Nutrients once locked in stone became available to ecosystems. The continents began to green.

Then came the forests.

By the Devonian Period, 390 to 360 million years ago, Earth experienced a botanical takeover. Towering plants with woody trunks spread across floodplains and deltas. Ferns, lycophytes, and early gymnosperms formed the first true canopies, their fronds casting shadows where only naked stone had stood. These forests were primitive in form but immense in impact. They altered river systems, changed erosion patterns, and reshaped continental hydrology. But their greatest legacy lay in what they built—and what they could not yet destroy.

The biochemical complexity of lignin—so essential for structural support—was beyond the digestive abilities of contemporaneous fungi and bacteria. No organism on Earth has yet possessed the enzymatic machinery to break lignin apart efficiently. As a result, when these early trees died, they did not fully decay. Their trunks fell into swamps and wetlands, accumulating layer upon layer of organic material. In an oxygen-poor environment, this material resisted decomposition. Instead, it thickened into peat—vast, dark, carbon-rich peatlands stretching across warm equatorial regions.

This slow burial became a planetary phenomenon. When Pangea began forming in the late Paleozoic, much of its equatorial belt was blanketed in colossal swamp forests that persisted for tens of millions of years. These forests drew carbon dioxide from the atmosphere, fixed it into biomass, and then, when they fell, locked that carbon underground. Over the next 60 million years, Earth performed the single greatest act of carbon sequestration in its history.

Pressure and time turned peat into lignite, lignite into bituminous coal, and bituminous coal into anthracite—layered seams of carbon hundreds of meters thick, running for thousands of kilometers beneath what are now North America, Europe, Russia, China, India, Australia, and South Africa. These seams were not local anomalies. They were planetary archives of photosynthesis—ancient sunlight pressed into stone.

But the forests did more than bury carbon. They exhaled oxygen— oceans of it. Atmospheric oxygen rose to levels as high as 30 to 35 percent, a concentration the modern world has not seen for hundreds of millions of years. The consequences were astonishing: insects grew gigantic, unconstrained by the diffusion limits that normally throttle their size. Dragonflies soared on meter-wide wings. Millipedes, the length of humans crawled through leaf litter deep enough to swallow a leg.

Yet high oxygen came with new dangers. The atmosphere became a tinderbox; lightning ignited wildfires with ease. The Carboniferous was a world of both flame and forest, a reminder that oxygen gives and oxygen takes away. Still, even fire could not keep up with the burial rates. Too many forests grew, too many trunks fell, too much carbon was entombed faster than it could return to the sky.

This period was not merely a chapter in Earth's natural history. It was a moment of planetary transformation. Life had never before altered the carbon cycle so profoundly, nor would it ever do so again with the same magnitude. The coal swamps of the Carboniferous did not simply shape the future—they became the fuel of futures. These ancient forests locked away energy that would remain dormant for hundreds of millions of years, waiting for a species that could dig, burn, refine, and repurpose it into engines, electricity, and a global civilization.

This carbonization was not inevitable. It required a perfect chain of contingencies: the evolution of lignin, the absence of lignin-degrading fungi for tens of millions of years, the presence of vast equatorial wetlands, the right temperatures, the right oxygen levels, and continents positioned to trap organic matter rather than disperse it.

It happened once.

Only once, in the deep history of the planet.

Forests were not merely ecosystems. They were planetary carbon vaults—massive, buried time capsules of stored starlight. In the quiet work of photosynthesis, growth, decay, and burial, Earth prepared the energy reservoirs that future species would treat as destiny.

In these ancient forests, the planet built the foundation for everything that would ever be powered, crafted, constructed, or dreamed by beings with hands, tools, and ambition.

The continents had awakened.

And the Earth, unknowingly, was beginning to save for the future.

CHAPTER 8

FOSSIL FUELS: ANCIENT SUNSHINE IN GEOLOGICAL STORAGE

How oceans, microorganisms, pressure, and deep time created the concentrated energy reserves that made civilization possible.

Long before humans walked upright, long before mammals warmed themselves under the canopies of ancient forests, long before the continents assumed their familiar shapes, Earth was quietly storing energy. It did not do so intentionally. It did not do so evenly. And it did not don so forever. But through a rare convergence of biology, chemistry, and geology, the planet locked away the sunlight of millions upon millions of years—saved in deep, pressurized vaults as if waiting for a species that could unlock them.

These are the fossil fuels: coal, oil, and natural gas—sunlight turned to matter, matter buried into stone, stone transformed by heat and pressure into reservoirs of usable energy. They are not accidents of nature. They are consequences of extremely unlikely planetary events.

Coal began its journey in the Carboniferous swamps, where lignin-rich forests refused to decompose. But oil and natural gas trace their ancestry to a different world altogether—one beneath the waves.

For hundreds of millions of years, the surface of the oceans was a drifting continent of microscopic life. Phytoplankton—tiny, fragile, exquisitely efficient—captured sunlight using the same photosynthetic machinery found in chloroplasts, themselves descendants of ancient cyanobacteria. Each cell was a bead of solar energy, pulled out of the sky and fixed into carbon bonds. When these organisms died, they sank. Most were consumed on the way down. But a fraction—just a fraction—reached the seafloor intact.

Where the bottom waters were anoxic, decay slowed. Organic matter accumulated. Layer upon layer of microscopic remains built up across vast continental margins, forming what geologists call sapropel, a soft, black, greasy mud rich in carbon. Over millions of years, more sediments piled on top, burying the organic matter deeper into the crust. Pressures rose. Temperatures increased. Chemical bonds rearranged. Kerogen—an intermediate, waxy substance—formed from the remains of marine life.

With further burial, kerogen cracked. Oil separated from the rock. Gas formed as temperatures climbed. The hydrocarbons migrated through porous sediments until they encountered traps—dome-like structures, faults, or impermeable layers that held them in place. These reservoirs accumulated across the ancient seabeds of the world, storing the liquefied history of countless generations of plankton and bacteria.

A barrel of oil is not a fluid.

It is a biography.

A million billion microorganisms lived, died, were buried, cooked, compressed, and preserved in the crust. And their chemical energy—energy first drawn from the Sun—remained locked away for hundreds of millions of years.

Coal and oil are different chapters of the same story: photosynthesis made storable.

This is why every hydrocarbon is a fossil.

Not figuratively.

Literally.

Molecular fossils of life.

Steranes, hopanes, porphyrins, and biomarkers still visible under mass spectrometry prove that petroleum is the filtered essence of ancient ecosystems. The oils of the Persian Gulf come from marine blooms that thrived in the Jurassic seas. The coals of Appalachia formed in equatorial wetlands before the breakup of Pangea. The gas fields of Siberia were once the remains of plankton drifting in the Permian sunlight.

These reservoirs formed only under precise conditions: high biological productivity, rapid burial, low oxygen levels, stable basins, and a just-right geothermal gradient. It is no wonder that, in the entire known solar system, Earth alone possesses these vast stores of energy. On most planets, carbon never concentrates. It oxidizes. It disperses. It cycles without ever accumulating into usable reserves.

Earth hoarded energy for a future it could not predict.

But this stored sunlight did more than wait. It shaped climate, regulated atmospheric composition, and stabilized global temperatures. It set the stage for ice ages and warm periods, extinctions and recoveries, as well as feedbacks and equilibria. The carbon cycle is not merely a chemical loop—it is the thermostat of a living world.

And yet the deepest consequence of this geological storage was latent. Dormant. Unrealized until a species with curiosity, manipulation, and ambition appeared—an animal that learned to dig, refine, burn, and command fire at industrial scale.

We are that species.

When humans tapped these reservoirs—when we mined coal, drilled for oil, and extracted natural gas—we unleashed hundreds of millions of years of stored solar power in a geological instant. The energy density of fossil fuels is not magical; it is historical. It is the product of ancient biospheres compacted into portable form. One gallon of gasoline contains the equivalent labor of tens of thousands of hours of plant photosynthesis. One ton of coal embodies centuries of forest growth.

Civilization was built on this inheritance.

Metallurgy. Transportation. Agriculture. Electricity. Cities. Technology. Industry.

Everything rides on the backs of ancient algae, ancient trees, ancient swamps, and ancient sunlight.

But this inheritance came at a cost: the carbon that had once been buried had to return to the sky. The atmospheric composition that forests spent millions of years regulating changed in a matter of decades. Humanity burned the geological library of life's past to power its present.

Yet the lesson is not that fossil fuels were a mistake.

The lesson is that they were a bridge.

A bridge from biological scarcity to technological abundance.

A bridge from muscle power to industrial power.

A bridge from the energy-constraints of the past to the energy-demands of the future.

A bridge from what evolution gave us to what physics demands of us.

But like all bridges, it cannot be the destination.

Because once a species awakens the energy of the past, it must learn to command the energy of the stars.

Civilization now stands at the threshold of the next leap.

Everything up to this point—photosynthesis, forests, carbon burial, hydrocarbons, combustion engines—has been preparation for the realization that energy is destiny.

Fossil fuels were Earth's savings account.

Fusion is the universe's income.

We used the past to build the present.

We must use the stars to build the future.

CHAPTER 9

THE MALTHUSIAN PHOTOSYNTHETIC TRAP

The long era in which humanity lived at the mercy of sunlight, muscle, and the thin trickle of energy that the biosphere could spare.

For all the brilliance of our species, for all our capacity to imagine, craft tools, and tell stories that spanned generations, humans lived for nearly all of their history inside a narrow energetic prison. The walls of that prison were invisible, but they were absolute. They were woven from the slow chemistry of photosynthesis and fixed in place by the limitations of muscle, wood, wind, and grain.

No matter how clever early humans became, they were bound by the same constraint that shaped every organism before them: the Sun only gives so much. Photosynthesis captures only a fraction of that light. Biomass grows only so fast. And everything alive—plants, animals, people—must share that limited harvest.

This was the photosynthetic trap.

Long before fossil fuels, before coal, oil, and turbines, humans lived on the energy that green plants could gather, store, and offer for consumption. That meant calories from crops, heat from wood, power from animals, and the occasional push of wind or water. These were not merely tools; they were ceilings. Their limitations defined the boundaries of civilization.

For tens of thousands of years, hunter–gatherers lived within the most basic version of the trap. They followed herds, harvested nuts and berries, and burned wood to cook food and illuminate their way through the night. Their world was one of abundance and scarcity in cycles, dictated by seasons and the migrations of prey. Human bodies—powered by roughly 100 watts of metabolic energy—were both the engines and the cargo of society.

When agriculture emerged around 10,000 years ago, it initially seemed like an escape. It allowed people to gather sunlight more efficiently through managed fields, enabling population growth and the rise of villages, which eventually led to the development of cities. But agriculture merely tightened the trap. Instead of living within the fluctuating productivity of ecosystems, humanity became tied to the fragile yields of staple crops—wheat, rice, millet, maize—each confined by the unchangeable limits of rainfall, soil nutrients, and the relatively slow energy output of photosynthesis.

Every calorie eaten began as a photon caught by a leaf.

Every human life depended on the energy trickle of the biosphere.

And that trickle was never enough.

For most of recorded history, civilizations hovered at the edge of famine. A few cloudy months, a drought, a locust swarm, or a harsh winter could cut harvests in half. Populations swelled during good years, pressing against the land's productive capacity, and then collapsed when the thin margin of surplus evaporated. This was the rhythm of the world: oscillation between growth and starvation, always constrained by the same simple reality—there was only so much sunlight falling on the fields.

Even fire, humanity's earliest tool for transforming energy, could not break the trap. Fire allowed cooking, warmth, pottery, metallurgy, and the forging of weapons. It extended the day and opened new ecological niches. However, in the pre-industrial world, fire was limited to what woodlands could supply. Forests grow slowly, and wood yields only a modest amount of heat per kilogram—barely enough for small furnaces, and certainly not enough to drive engines or shape steel at industrial scale without exhausting the land.

By the Middle Ages, entire regions of Europe were deforested, not because of malice or ignorance, but because societies had reached the inherent limit of biomass. Wood was energy, and energy was life, and there was never quite enough of it.

Animals added power, but only a little. Oxen and horses could plow fields or haul loads, but their muscles ran on the same photosynthetic bottleneck. Every draft animal consumed food that could have fed people. Every increase in horsepower required an increase in cropland. The trap tightened with each attempt to push beyond it.

Windmills and waterwheels offered relief, but their power was diffuse and unpredictable—useful for grinding grain or sawing timber, but incapable of driving societies beyond subsistence. They skimmed the surface of natural flows but never altered the deeper arithmetic of energy.

Civilizations rose and fell within these constraints. Rome, Egypt, Mesopotamia, the great dynasties of China, the civilizations of the Indus Valley and the Andes—all achieved remarkable complexity, each stretching the photosynthetic economy to its breaking point. But none escaped it. Their populations expanded until they reached the limits of agricultural yields, and then stagnated or declined. Cultural achievements soared, but energetic ceilings held firm.

Even after the Renaissance and Scientific Revolution, even as telescopes scanned the heavens and printing presses multiplied ideas, Europe still heated its homes with wood, illuminated its streets with tallow or whale oil, and powered its industries with human sweat and animal muscle. Knowledge grew, but energy did not, and so humanity remained bound to the slow, green chemistry of leaves.

This is the crucial truth:

Intelligence alone does not liberate a species from energy poverty.

Only access to new energy sources does.

For our ancestors, the world was a mosaic of hard limits—on land, food, fuel, and labor. Progress could only be incremental, fragile, and reversible. The brightest minds of antiquity lived in a world where the total energy budget of a city—tens of thousands of people—was less than that consumed by a single modern passenger jet taking off.

The trap was not cultural.

It was not moral.

It was not political.

It was physical.

Every human civilization before the eighteenth century was confined to the narrow trickle of photosynthesis, where the thin beam of sunlight was converted into grain, wood, and grazing land. The biosphere rationed its energy output, and the species rationed its ambitions.

And yet, beneath the soil and sea, the planet held a secret.

Unknown to the ancient world, Earth had spent hundreds of millions of years storing the sunlight of vanished forests and extinct plankton. It had locked away the chemical energy of Carboniferous swamps and Mesozoic seas in seams of black rock and reservoirs of dark, viscous oil. It had buried the excesses of past biospheres with the slow patience of deep time.

Humanity lived in energy poverty while sitting atop unimaginable wealth.

But it would take fire—refined, disciplined, weaponized fire—to unlock it.

And when humans discovered how to burn not just wood but the ancient sunlight buried in coal and oil, the trap shattered forever.

CHAPTER 10

FIRE: HUMANITY'S FIRST ENERGY REVOLUTION

How the control of flame reshaped human evolution, metabolism, culture, and destiny — the moment an animal first reached beyond the limits of its own body.

Fire was not humanity's invention.

It burned on Earth long before our ancestors appeared — lightning strikes ignited grasslands, volcanic vents lit forests, and wildfires swept across landscapes in glowing, roaring waves. Fire was ancient, fierce, and untamed. But for most of Earth's history, it belonged to the weather, to geology, to the blind choreography of atmospheric oxygen and dry vegetation.

No creature commanded it.

No organism understood it.

No lineage shaped its future.

Until humans.

Somewhere in Africa, over a million years ago — perhaps as early as 1.7 million years — hominins gathered at the edge of a wildfire, drawn to the heat and the strange softness of cooked meat left behind. The first interactions were accidental, opportunistic, and timid. But slowly, over generations, early humans learned not just to approach fire, but to carry it, nurture it, and protect it.

When our ancestors learned to keep a flame alive — banked in embers wrapped in bark or hidden under ash — everything changed.

For fire is not merely heat.

It is external metabolism.

The human body operates on a chemistry so complex that it barely sustains its own temperature. But fire burns at thousands of degrees, releases energy at rates no muscle could match, breaks down organic materials more efficiently than teeth or digestive enzymes, and transforms the environment with a ferocity evolution had never seen.

When early humans brought fire under control, they acquired a second stomach, a second furnace, a second source of power — outside the body but serving its needs. As Richard Wrangham proposed, cooking may have been the catalytic event that made modern humans possible. Cooked food is softer, richer in available nutrients, and easier to digest. It allowed our ancestors to extract far more energy from the same calories, freeing metabolic resources for the expansion of the brain.

With fire, our species grew smarter.

But fire also made us safer.

Night ceased to be the domain of predators.

Caves became homes.

Savannas became kitchens.

Communities gathered around flames, and language found its earliest fertile ground in the circle of flickering light.

But perhaps most importantly, fire taught humans the first lesson in energy discipline.

It taught us to gather fuel, to ration heat, to plan ahead, to understand the relationship between fuel quality, flame intensity, and time. Long before humans developed mathematics or writing, they were already students of thermodynamics.

Yet fire remained bound to biomass.

Its fuel was wood, brush, and dung — the meager energy stores of living ecosystems. For hundreds of thousands of years, human populations remained small because energy remained scarce. A tribe's size was limited by the productivity of its landscape. Too many people meant too many fires, too much fuel gathered, too little forest left to burn.

Even as Homo sapiens spread across Africa, then to the

Middle East, Europe, Asia, Australia, and eventually the Americas, they carried the same constraint: fire needed wood, and wood needed time. Whole ecosystems had to recover after being stripped for heat. The photosynthetic trap, though widened by human ingenuity, still held.

But fire was the first crack in the wall.

Fire allowed the forging of stone tools, then bone and antler, and eventually — millennia later — bronze and iron. Metallurgy is not a biological process. It is a harnessing of temperatures that evolution never intended biological organisms to command. The moment humans smelted ore, they had crossed another threshold: from survival to transformation.

Civilizations rose with fire as their foundation.

But again, they hit limits.

Wood is abundant but diffuse. A mature tree represents only decades of energy accumulation — and when burned, its heat is quickly spent. No matter how skilled the blacksmith, no matter how advanced the furnace, charcoal and wood could not generate enough sustained heat to smelt the high-quality iron needed for machines, railways, engines, or heavy industry. They could sustain craft, but not progress.

For tens of thousands of years, humans lived at the edge of a paradox:

Fire gave us power, but only enough to desire more, not enough to achieve it.

Wind could sail ships, but not drive industry.

Water could turn wheels but not reinvent cities.

Animals could plow fields but not forge steel.

Muscle could build monuments, but not mass production.

Fire remained tethered to the biological world that made it possible.

And humanity remained tethered to the energy limits of the biosphere.

It would take another great leap — another liberation from the photosynthetic trap — for humanity to escape the ceiling of natural energy flows.

That leap came when fire met fossils.

When humans discovered coal, the compressed sunlight of ancient forests, and learned to burn it not just for warmth but for mechanical power, the energy math of the species changed forever. The steam engine transformed fire from a passive warmth into an active force — a controlled explosion driving pistons, turning wheels, and finally breaking the power bottleneck that had constrained civilization for all of recorded history.

But that story belongs to the next chapter.

For now, it is enough to understand this:

Fire was humanity's first assertion of independence from the limitations of biology.

It was our first partnership with energy at a scale beyond muscle.

It was our first glimpse of mastery over the forces that shape worlds.

When early humans encircled flame and kept it alive through the night, they were not just warming themselves.

They were rehearsing for the future.

The future in which they would burn forests… then peat… then coal… then oil… then atoms… and one day, the future in which they will ignite the stars.

COSMIC ORIGINS

Big Bang	13.8 billion years ago
Formation of fundamental particles	13.7 billion years ago
Hydrogen and helium emerge	13.6 billion years ago
First atoms form	13.4 billion years ago
First stars ignite	13.3 billion years ago
First galaxies form	12.4 billion years ago
Nearby supernovae create heavy elements	~4.3 billion years ago
Dust clouds and molecular ices catalyze chemical reactions	before 4.6 billion years ago
Water and organics delivered (comets and asteroids)	before 4.6 billion years ago
First organic molecules	before 4.6 billion years ago

SOLAR ORIGINS

Collapse of the solar nebula	4.6 billion years ago
Birth of the Sun	4.57 billion years ago
Formation of the planets	4.56 billion years ago
Protoplanetary collisions	4.51 billion years ago

CHEMICAL BEGINNINGS

Great Oxygenation Event	2.4 billion years ago
Snowball Earth cycles	2.3–0.7 billion years ago

CHEMICAL BEGINNINGS

Great Oxygenation Event	2.4 billion years ago
Snowball Earth cycles	2.3–0.7 billion years ago

HUMAN ASCENT

First hominins	6–2.5 million years ago
Stone tools	~3.3 million years ago
Mastery of fire	~300,000–1,000,000 years ago
Language	~130,000 years ago
Cultural explosion	50,000–40,000 years ago
Agriculture	~10,000 years ago
Cities and civilization	~6,000 years ago
Bronze Age	~5,000 years ago
Iron Age	~3,200 years ago

ENERGY REVOLUTIONS

Whale oil (illumination)	1700–1850 CE
Coal	1750–1900 CE
Electricity	1850–present
Oil and gas	1900–present
Abandonment of whale oil	1880s–present
Internal combustion engine	~1900–present
Nuclear power	1950s–present

ENERGY REVOLUTIONS

Whale oil (illumination)	1700–1850 CE
Coal	1750–1900 CE
Electricity	1850–present
Abandonment of whale oil	1880s–present
Internal combustion engine	~1900–present
Nuclear power	21st century and beyond

CHAPTER II

THE LIMITS OF WOOD, WIND, WATER, AND MUSCLE

Why every ancient civilization, no matter how brilliant, eventually ran into the same unbreakable limit.

For nearly all of human history, the world dimmed at sunset. Cities glowed not with electricity but with flickering flames—torches in the streets, whale oil in lamps, hearth fires in stone houses. The night belonged to shadows, to the hush between breaths, to a thin ribbon of radiance coaxed from burning wood. Across continents, from the Yellow River to the Nile, from the Tiber to the Andes, the energy of human civilization burned with the same small flame.

It was not ignorance that dimmed the ancient world.

It was physics.

Muscle, fire, wind, and flowing water were the only tools our ancestors possessed, and all four drew ultimately from the same source: sunlight captured by photosynthesis. This was the ceiling of the pre-industrial world. The Sun offered its energy freely, but in a form so diffuse and slow that even the brightest civilizations lived close to the edge of scarcity.

No society escaped this limit.

Not one.

The pyramids of Egypt rose through the coordinated labor of tens of thousands of workers, each powered by a diet whose calories began as grains pushed up through the soil by sunlight. The Roman aqueducts, masterpieces of stone and geometry, were constructed by human hands and animal muscle. The Great Wall of China stretched across mountains on the backs of soldiers and peasants who ate millet, rice, and barley—all watered by rains, all nourished by the thin trickle of solar energy stored in plants.

Every stone lifted by a slave or a laborer was the conversion of sunlight into motion.

Every cart pulled by an ox, every plow dragged by a horse, every war machine heaved across a battlefield was bound to the photosynthetic budget of its land.

Human brilliance could amplify energy, but it could not create it.

Rome learned that lesson slowly, and then all at once. Its expansion required ships—thousands of them—to connect the Mediterranean world. Each trireme devoured the forests of Italy, Greece, and Asia Minor, for a single warship could consume several acres' worth of trees. By the height of the Empire, the Italian peninsula was nearly stripped bare. Timber had to be imported from faraway lands simply to maintain the infrastructure of empire. The Roman economy did not collapse due to decadence; it collapsed due to deforestation.

China confronted the same constraint. Dynasties rose and fell in tandem with the rhythms of agriculture, harvests, and the energy embodied in grain. The great furnaces of the Han and Tang eras required wood charcoal, and charcoal in turn demanded forests. Iron production soared and then faltered as wood supplies dwindled. Even advanced metallurgy—something Europe would not match for a thousand years—could not break free from the slow regrowth of trees.

Japan faced the same fate. By the 1600s, under the Tokugawa shogunate, the country faced a full-blown energy crisis. Wood was used for heating, cooking, construction, shipbuilding, smelting, and the production of ceramics. Forests vanished with alarming speed. Entire provinces stood on the edge of ecological collapse. Only through extreme laws, reforestation programs, and rationing did Japan avoid disaster—but it did not escape the fundamental constraint. It merely postponed it.

Greece, once forested, became the rocky, scrub-covered landscape we know today because its classical builders and its navy consumed trees faster than they could grow. The Maya cleared jungles for limestone kilns, their cities rising with white brilliance only to collapse under the weight of their own energy needs. The Khmer Empire, builders of Angkor's monumental temples, consumed immense resources for agriculture and stone transport, but even their hydraulic genius could not overcome the metabolic ceiling of a biomass economy.

Wind and water helped—but only at the margins.

A river could turn a wheel, but it could not smelt iron.

A sail could cross an ocean, but it could not power a city.

A windmill could grind grain, but it could not lift civilization out of the photosynthetic trap.

Every ancient innovation worked within the same narrow budget of available energy—energy gathered by plants and stored temporarily in wood, grain, and muscle. The power of a society rose and fell with the harvest. A bad season meant hunger. A series of poor seasons led to collapse.

For all their insight, ingenuity, and architectural vision, the ancients were trapped by energy constraints they could not see but always felt. They could understand geometry well enough to build pyramids, hydraulics well enough to construct aqueducts, astronomy well enough to predict eclipses—and yet none of this brilliance improved the fundamental energetic equation.

The ceiling remained.

Ruthless.

Unbroken.

Universal.

Even Isaac Newton — perhaps the greatest scientific mind ever to walk the Earth — lived under the same ceiling. Newton, the Lucasian Professor at Cambridge, Fellow of Trinity College, architect of classical mechanics, inventor of calculus, and the man who first revealed that apples and planets obey the same universal law, understood the heavens with unmatched clarity. He calculated escape velocities. He traced the orbits of comets. He could describe, in perfect mathematics, how one might leave the Earth. And yet Newton could never have gone to the Moon. Not because he lacked genius or imagination, but because he — and everyone else of his age — lacked the energy. To lift a vessel into space requires power so concentrated and explosively dense that the biomass economy of the pre-industrial world could not come close. Newton could chart the path to the Moon, but he could not take the first step. His world burned wood and muscle; ours burns the energy of ancient stars. Humanity did not reach the Moon because it became smarter, but because it became more energetic. This is the quiet truth that binds all civilizations: brilliance expands possibility, but only energy expands reality.

The largest cities of antiquity consumed wood from tens of miles around. The greatest empires strained their landscapes to the brink. Every incremental rise in population, construction, metallurgy, and transportation was accompanied by a corresponding increase in fuel consumption. And the forests

—slow-growing, limited by rainfall and soil—could not keep up.

It was not a failure of imagination.

It was a failure of supply.

The pre-industrial world was a zero-sum game. For one society to rise, another had to fall—or a new landscape had to be conquered and stripped for wood. Expansion bought time, but not freedom.

Civilizations advanced until they encountered a similar barrier, and then they stagnated.

Wood could not deliver the heat needed for higher-grade metals.

Charcoal could not sustain production at scale.

Animal power could not accelerate beyond biological limits.

Wind and water could not industrialize.

Fire, without new fuels, could not do more than humans already demanded of it.

For thousands of years, humanity lived with genius and ambition—but starvation and stagnation at its heels.

Yet beneath their feet, locked under their fields, under their mountains, under their seas, lay the energy of worlds.

Coal waited in the strata of the Carboniferous.

Oil waited in the sedimentary basins of vanished seas.

Gas slept in the pores of ancient stone.

Stored sunlight — millions upon millions of years of it — lay dormant.

The ancients never knew.

They had no reason to suspect it.

Their entire experience of energy was tied to trees, rivers, wind, and muscle.

But the Earth had prepared a different future.

And that future arrived the moment humans burned their first lump of coal.

CHAPTER 12

COAL: THE FIRST ESCAPE FROM THE TRAP

How a black stone forged in the swamps of deep time gave humanity the thrust needed to leave the biological limits of Earth.

Coal does not look like destiny.

It looks like darkness — a dull black stone veined through the crust of old continents, easily overlooked, easily dismissed. For most of human existence, coal lay unchanged beneath our feet while civilizations rose and fell above it, each straining against the same ceiling of muscle, wood, wind, and water.

But coal is not merely carbonized plant matter.

It is compacted sunlight — millions of seasons of photosynthesis, buried, compressed, heated, and transformed into energy far denser than anything the biosphere could produce in real time.

For thousands of years, humanity had lived off the thin, daily allowance granted by photosynthesis. But beneath the fields and forests, Earth had stored a secret:

The energy of millions of years, waiting in reserve.

Coal was Earth's first great gift to a species capable of using it.

Coal had been known since antiquity — the Greeks called it "anthrax," the Romans used it in minor quantities, and the Chinese extracted it for ornamental carving and small-scale heating — but none of these cultures recognized its true potential. Why would they? Their economies still revolved around wood, charcoal, animals, and the living pulse of the land. Coal was harder to ignite, dirtier to handle, and unnecessary for a world powered by forests.

Coal remained a curiosity because no civilization had yet built a machine hungry enough to need it.

That moment arrived in the late 17th and early 18th centuries, when England faced a familiar tragedy: forests stripped to their stumps. Timber shortages reached crisis levels. Shipbuilding, heating, metallurgy, and glassmaking all gnashed against the limits of the biosphere. The same ceiling that had capped Rome, China, and Japan now hung over Britain.

But Britain also had something else:

Easily accessible coal seams.

Coal mines deepened, water seeped in, and miners suffocated in tunnels unless the groundwater could be removed. The old tools — buckets, horses, windmills — couldn't keep the mines dry.

Out of this desperation came the first practical steam pump, designed by Thomas Newcomen in 1712. The engine was crude, inefficient, and massive — but it worked. It burned coal. And in doing so, it performed labor no horse, river, or human could match.

This was the first moment in history when a machine delivered more energy than a living organism.

The implications were staggering.

Suddenly, energy no longer depended on the speed of plant growth or the strength of muscle. It depended on how much coal could be extracted and burned. The equation of civilization changed forever.

Within a few decades, James Watt refined the steam engine, significantly boosting its efficiency and transforming it from a mining tool into the driving force behind industry. Factories filled with looms, forges, and pumps. Trains roared across continents. Ships crossed oceans without wind. Cities expanded beyond what wood could ever have sustained.

Coal turned fire from a heat source into a force multiplier.

The heat released from burning coal is not a mystical phenomenon. It is chemistry:

carbon atoms once bound into the trunks of prehistoric trees now recombine with oxygen, releasing the energy stored in their molecular bonds. But the scale is unmatched. A single ton of coal contains more usable energy than centuries of forest growth. The industrial world was no longer living on the sunlight of today — it was burning the sunlight of the Carboniferous.

Coal changed everything:

1. **Time scale** - Humanity moved from real-time energy production (annual crops) to geological-time energy production (millions of years condensed).
2. **Geographic scale** - Factories concentrated energy so intensely that production no longer needed to sprawl across landscapes. Cities became engines.
3. **Labor scale** - A coal-fired machine could do the work of hundreds of men. It was as if human civilization suddenly discovered extra limbs — whole mechanical musculatures — with no additional biological cost.
4. **Speed scale** - Where ancient societies grew slowly, limited by timber and seasons, industrial societies grew explosively, limited only by coal seams and human imagination.

5. **Life scale** - Life expectancy rose. Disease declined. Nutrition improved. Transportation revolutionized finances, trade, and ideas.

Coal was the first true escape velocity of human society — the moment we achieved more power than biology alone could ever provide.

But coal did more than free humanity from the photosynthetic trap.

It launched the first global energy acceleration — a curve that has continued to climb.

It powered the iron foundries that built railroads and steamships.

It drove the factories that poured steel, spun textiles, and forged machines.

It lit the gas lamps of cities and heated the boilers that became the heart of modern industry.

For the first time in Earth's history, a species was no longer constrained by the power of its own muscles or the growth rates of forests.

Coal expanded the human horizon.

Coal unlocked the industrial revolution.

Coal mapped the world into time zones, connected continents by telegraph, and reduced the time it took to cross the oceans from months to days.

And yet, as we now know, coal carried a long-term cost:

each pound burned returned Carboniferous sunlight to the atmosphere.

What forests had buried over sixty million years, humanity burned in a few centuries.

But that reckoning belongs to later chapters.

For now, the story is this:

Coal was the first time humanity tasted abundance — real, accelerating, unprecedented abundance.

It was the first time our species stepped beyond the natural limits evolution had assigned us.

Coal turned fire into power.

Power into industry.

Industry into civilization.

And civilization into something that could one day look beyond Earth itself.

From here, the story accelerates.

Oil waits.

Electricity waits.

Nuclear fission waits.

And beyond all of them — fusion, the energy of the stars.

Coal was the first rung.

But humanity is climbing toward something far brighter.

CHAPTER 13

PETROLEUM: THE LIQUID MIRACLE

How a viscous black fluid formed in ancient seas became the indispensable bloodstream of modern civilization.

Crude oil does not look like the engine of a civilization. It is thick, dark, and brooding—an ancient broth of once-living things transformed by heat, pressure, and time. It pools in quiet basins beneath the Earth, gathering in traps formed by tectonics and the long choreography of deep time. For millions of years, it lay buried and inert, waiting for a species clever enough, desperate enough, and ambitious enough to find it.

Humanity stumbled into that discovery almost by accident.

For centuries, people had known of "rock oil"—seepages that oozed from the ground, used in small quantities for medicines, lubricants, and waterproofing. But these were curiosities, not the lifeblood of an energy economy. No civilization before the 19th century had reason to imagine that the black liquid beneath their feet was the most concentrated, portable, versatile store of chemical energy the planet had ever created.

The world still ran on coal, steam, and muscle.

Then came Titusville, Pennsylvania, in 1859.

A driller named Edwin Drake sank an iron pipe into the ground in search of oil. When petroleum bubbled up through his well, it was not just a discovery—it was a detonation in the history of energy. That moment marked the birth of the modern world. In one stroke, humanity transitioned from the dense, heavy, infrastructure-bound fuel of coal to a substance so energy-rich and so easy to move that it rewrote the rules of civilization.

Coal had given us the industrial age.

Oil gave us the age of motion.

Oil is a miracle not because of magic, but because of physics. Liquids flow. Liquids can be pumped, stored, transported, refined, fractionated. And the hydrocarbons within crude oil—long chains of carbon and hydrogen forged from the remains of ancient plankton—carry an energy density unmatched by any fuel humanity had ever known.

A single gallon of gasoline contains the work equivalent of hundreds of hours of human labor. No draft animal, no waterwheel, no windmill, no steam engine burning coal could match the raw, explosive power of refined petroleum.

Suddenly, machines no longer needed to be tethered to the coal mine or railroad. They could move. They could roam. They could accelerate.

The internal combustion engine was not merely a technological advance; it was the unlocking of a new energetic regime. Instead of boiling water to push pistons, engines could burn vaporized fuel directly within their cylinders. The result was unparalleled energy conversion—violent, precise, controlled detonations happening thousands of times per minute. This unleashed a degree of power previously unimaginable.

The world transformed almost overnight.

Cars appeared first as novelties, then as necessities.

Roads multiplied.

Families gained mobility.

Trade expanded inland.

Cities sprawled.

Distances shrank.

And so did oceans.

Petroleum fueled the great steamships that crossed the Atlantic, then the fleets of the First World War, then the aircraft that stitched continents together. Aviation—impossible without high-energy liquid fuel—became the new frontier of human movement. It carried mail, people, ideas, armies, and cultures across the sky at speeds faster than any organism had ever traveled.

But oil's genius did not stop at transportation.

Refineries revealed that crude oil was not a single substance but a library of molecular possibilities. Gasoline, diesel, kerosene, jet fuel, naphtha, lubricants, solvents, and asphalt —each distilled from ancient sunlight, each powering a different sector of civilization. And then came something even more consequential: plastics.

From polyethylene to nylon to PVC to polystyrene, petrochemicals created an entire artificial biosphere—light, strong, sterile, flexible, indispensable. Modern medicine, agriculture, communication, electronics, textiles, and global trade rely on materials that did not exist before the refinement of crude oil.

The modern world is not merely built with oil. It is made of oil.

And still, petroleum's influence reached deeper. It reshaped geopolitics, drawing borders in deserts and oceans. It forged alliances and provoked wars. It created wealth and inequality on planetary scales. It allowed the global population to grow from 1 billion to 8 billion in just two centuries by powering tractors, fertilizers, pipelines, power plants, trucks, and grid infrastructure. It fed billions by fueling the Haber–Bosch process, which creates ammonia—the foundation of synthetic fertilizers, and by extension, modern agriculture.

Petroleum became the bloodstream of a planet-wide civilization.

And yet, like coal, oil was a finite inheritance. Its abundance was a consequence of ancient oceans where planktonic life bloomed in staggering quantities, died, sank, and was buried under anoxic conditions. It required geological luck, tectonic basins, and millions of years of heat at the right depths. A thousand planets could have oceans and never produce a drop of petroleum. A thousand intelligent species could search for it and never find the reservoirs Earth gifted to us.

Oil is rare.

Oil is ancient.

Oil is irreplaceable due to its unique combination of density, portability, and chemical versatility.

And it arrived at the exact moment humanity needed it to escape the limits of coal. Without petroleum, there are no cars, no aviation, no container ships, no globalized world. There is no 20th century as we know it.

But petroleum, like coal, was only a chapter—

the second great escape from the photosynthetic trap.

Because once humanity learned to command liquid fuels, its ambitions grew beyond transportation and industry. It sought cleaner combustion, greater efficiency, higher temperatures, faster turbines, and new horizons in chemistry and power generation.

What followed was natural gas.

Then fission.

And beyond that, the still-unrealized promise of fusion.

Petroleum was the liquid miracle.

But it was not the destination.

It was the engine of the modern world—and the bridge to the next.

The First Encounters With Oil

Long before the world ran on petroleum, long before cities glowed and engines roared, humans encountered oil in its rawest, most primitive form. It seeped from the ground in quiet places—on riverbanks, in marshes, in lowland basins—and it offered more nuisance than opportunity. In ancient Mesopotamia, bitumen oozed from the earth, a sticky, dark substance. It fouled wells, coated reeds, and clung to the bottoms of sandals. People used it sparingly as a waterproofing agent or as mortar for bricks, but no one imagined that this viscous fluid would one day power global civilization.

In ancient China, drillers searching for brine sometimes struck oil instead. The strange liquid ruined salt production and contaminated water. There are accounts of wells in medieval Europe that were so thoroughly contaminated by petroleum that they had to be abandoned altogether. Oil was, in the ancient world, not a resource but an intrusion.

Even when it was deliberately collected, its uses were few. Lamps could burn it, but the results were smoky and foul-smelling. It behaved poorly compared to whale oil—the clean, bright-burning lubricant that lit the cities of the 18th and early 19th centuries.

For a time, whales paid the price.

The whaling fleets of New England ranged far across the oceans, harvesting sperm whales, right whales, and bowheads for oil. Whale oil filled lamps, lubricated machinery, and provided light for reading, sewing, commerce, and science. It was precious, expensive, and limited. And whaling—dangerous, brutal, and ecologically disastrous—could not continue indefinitely. Some whale populations collapsed entirely. Others hovered near extinction.

Petroleum saved them.

When kerosene—cleaner, brighter, and vastly more abundant—entered the market in the mid-1800s, the demand for whale oil collapsed almost overnight. Whaling ports emptied. Ships rotted on docks. Entire industries vanished. Petroleum, for all its imperfections, rescued whales at the very moment they were disappearing from the oceans. It was one of the great, unintended mercies of technological evolution.

But the greatest transformation petroleum offered was not in lighting—it was in motion.

The internal combustion engine, perfected in the late 19th

century, unleashed a revolution that steam never could. A steam engine is a heavy, thirsty creature, requiring enormous boilers, vast amounts of fuel, and constant maintenance. It is shackled to infrastructure—rails, mines, ports. But a combustion engine is small, light, portable. It burns its fuel inside its cylinders—controlled detonations delivering sudden, explosive power.

Steam engines transformed industry.

Internal combustion engines transformed geography.

Cars spread first as curiosities, then as tools, and finally as necessities. Tractors liberated fields from centuries of backbreaking labor. Trucks liberated cities from dependence on canals and railroads. Planes liberated entire nations from isolation. And each of these machines ran on petroleum—a fuel that did not need a furnace, a stoker, a boiler, or a crew. A small tank could power hours of travel, days of labor, lifetimes of mobility.

Petroleum did not just accelerate civilization.

It unlocked it.

And yet the distribution of oil across Earth's crust was never even. It was the product of ancient seas, tectonic basins, and geological luck, not politics or morality. The location of oil shaped the geopolitics of the modern world, yes—but oil itself was never to blame for how people used it. It is a neutral inheritance, a gift shaped by deep time, its impacts determined by human choices, not by its chemistry.

The moral dimensions of energy—its stewardship, responsibility, and future—will be explored in later chapters. When we reach the point of fission and fusion, when we discuss the new imperative, that is the moment to explore the ethical terrain of energy and destiny.

For now, the story of petroleum remains this:

a. Oil was a curiosity.
b. Then a nuisance.
c. Then a lamp fuel.
d. Then a miracle.
e. Then the bloodstream of modern civilization.

The discovery of petroleum did not just change how humanity moved.

It changed what humanity was capable of becoming.

CHAPTER 14

NATURAL GAS: THE CLEANEST BRIDGE

How methane, the simplest hydrocarbon in the universe, became the transparent flame illuminating humanity's transition toward a nuclear future.

Natural gas is nearly invisible when it burns. Its flame is blue, quiet, almost shy — a soft cone of light rather than the roaring orange of coal or the smoky flicker of wood. Yet within that quiet blue flame lies one of the most extraordinary energy gifts Earth ever produced: a fuel so clean, so powerful, and so chemically elegant that it now undergirds the entire modern world.

Methane is simplicity incarnate — one carbon atom embracing four hydrogens, a structure so basic it forms spontaneously in outer space, in the atmospheres of gas giants, in the cryovolcanic plumes of icy moons, and even in the interstellar dust clouds where stars are born. But on

Earth, methane's meaning is different. Here, it is the gaseous remnant of life: ancient marine plankton, terrestrial vegetation, microbial action in sediments — all transformed through burial, pressure, and time into a gas that burns with almost mathematical cleanliness.

Where oil is a dark, complex cocktail of long hydrocarbon chains, methane is pure geometry — a tetrahedral molecule storing energy with exquisite efficiency. When it burns, the chemistry is astonishingly tidy:

$$CH_4 + 2\ O_2 \rightarrow CO_2 + 2\ H_2O + \text{energy}$$

Nearly perfect combustion, with remarkably little soot, ash, residue, or toxic byproducts. Every spark is a small masterpiece of thermodynamic elegance.

For this reason, methane has become the backbone of modern energy.

The Invisible Engine of the Present

Natural gas is so ubiquitous today that it has become almost invisible. It heats homes, powers stoves, dries grain, fires industrial furnaces, and runs the boilers of modern districts. But its most transformative role is in electricity — in the rise of the modern gas turbine.

The gas turbine is one of the greatest engineering achievements of the 20th century. Compress air. Inject methane. Ignite. Expand the hot gases across turbine blades spinning at tens of thousands of revolutions per minute. Convert that motion into electricity with astonishing efficiency. Modern combined-cycle plants — where waste heat is harvested to produce additional steam power — reach efficiencies above 60%, the highest of any fossil-fuel technology ever created.

Methane produces more usable energy per molecule of CO_2 than any fossil fuel before it. Its lack of soot is not cosmetic; it is a chemical virtue. In a world still tied to hydrocarbons, methane is the cleanest form of combustion available at scale.

And unlike solar and wind, gas turbines provide the dependable backbone that keeps hospitals lit, data centers running, grid frequency stable, and cities alive when the wind falters or the night stretches on.

Methane is not the future.

But it is the bridge to the future.

Without it, the transition to fission would collapse under the weight of intermittency and fragility.

The Rise of Gas in the Industrial World

For much of the early petroleum era, natural gas was considered a nuisance — a dangerous, explosive byproduct encountered during drilling. Oil men flared it uselessly into the sky, burning rivers of fire above derricks. Entire fields glowed at night with the flames of wasted methane, the atmosphere shimmering with heat from energy too difficult to transport.

But pipelines changed everything.

Once engineers learned to compress, liquefy, store, and transport methane, the world entered a new phase of energy abundance. Cities lit with natural gas long before electricity grids matured. Factories converted furnaces to cleaner flames. Homes shed the soot of coal-fired stoves. And agriculture — the quiet miracle of industrial food — surged into a new era. The Haber–Bosch process, which synthesizes ammonia from nitrogen and hydrogen, depends on natural gas as both feedstock and fuel. Nearly half of the nitrogen atoms in human bodies today originate from natural gas.

Methane has become the thermodynamic scaffolding of civilization's present moment.

Gas Turbines: Humanity's Most Elegant Flame

Gas turbines deserve a place alongside the steam engine and the internal combustion engine as pillars of technological transformation. Their brilliance lies in their internal simplicity and external sophistication: a machine that inhales air, compresses it, injects methane, ignites it, and spins out enormous power.

Where steam engines required mountains of coal and armies of workers to feed boilers, gas turbines require only air, fuel, and engineering. Their compactness, efficiency, and low emissions have made them the quiet giants of electricity generation.

And perhaps most importantly, gas turbines provide the stability and flexibility that grids need to coexist with renewable energy. They can ramp up in minutes, compensate for dips in wind or solar output, and maintain the reliability without which modern civilization would snap.

The Perfect Bridge — and Why Bridges Matter

There is a tendency among some to view natural gas as merely another fossil fuel, indistinguishable from coal or oil. But this is a categorical mistake — a misunderstanding of chemistry, engineering, and the thermodynamic arc of civilization.

Methane is not the destination.

But it is the only realistic bridge that can support the world while we scale fission and develop fusion.

Coal takes decades to phase out.

Oil will remain indispensable for aviation and petrochemistry.

Renewables cannot yet provide baseload stability or industrial heat at scale.

Fission is essential, but it is slow to build.

Fusion is inevitable but not imminent.

In this landscape, natural gas is the right-now solution — abundant, clean-burning, efficient, and technologically mature. It is the literal fuel of the transition: a stabilizer while we construct the next chapter of human destiny.

A Hydrocarbon of Responsibility, Not Regret

Natural gas is sometimes framed in moral or political terms, but such framing misses the point. Energy is not a moral agent. It is a resource. A tool. A physical inheritance shaped by time, geology, and chemistry. The ethics arise not from the molecule, but from how we use it.

And methane, more than any other hydrocarbon, offers humanity the chance to move forward responsibly while buying the time we need to build reactors, develop new materials, perfect safety, establish regulatory frameworks, and prepare for the technological leap that will define the next century.

Natural gas is not the villain of the story.

It is the bridge that carries us from one world to another.

From coal's smoke to oil's motion.

From oil's motion to methane's clarity.

And from methane's clarity to the controlled fire of the atom — and ultimately, to the bottled starfire of fusion.

This is the cleanest bridge.

And it is the necessary one.

CHAPTER 15

HYDROGEN: THE MISUNDERSTOOD NON-FUEL

Why the universe's simplest atom is not the energy source people imagine — and what batteries, electrons, and hybrids really mean for a civilization hungry for power.

Hydrogen is the most abundant element in the universe. It fills the stars, drifts in clouds between galaxies, coats the icy moons of the outer planets, and forms the building block from which the cosmos assembled itself. Every atom heavier than helium was forged in stars, but hydrogen was here from the start — a primordial simplicity.

Perhaps it is this cosmic ubiquity that seduces people into believing hydrogen is an energy solution. If the universe runs on hydrogen fusion, they reason, then surely humanity can run on hydrogen too.

But this is a misunderstanding of staggering proportions.

The universe runs on fusion, not hydrogen.

Hydrogen on Earth does not fuse.

Hydrogen on Earth does not exist in free form.

Hydrogen on Earth is not a fuel at all.

It is merely a battery — a storage medium — dependent entirely on energy that must be supplied from somewhere else.

And the "somewhere else" is almost always natural gas, coal, oil, hydroelectricity, or nuclear fission.

This is the part people fail to understand.

Hydrogen is not an energy source.

Hydrogen is an energy carrier — and a rather inefficient one.

The Mirage of Hydrogen

Hydrogen does not exist freely on Earth because it is too light to be held by gravity. Any molecule that breaks free into the atmosphere quickly escapes into space. This alone makes hydrogen a poor terrestrial fuel: the planet simply does not have any to harvest.

Every molecule of hydrogen used in industry — every gram for fuel cells, for "green hydrogen," for ammonia synthesis, for metallurgy, for rocket fuel — must be manufactured by humans through an energy-intensive process. We can split water (an expensive, inefficient process), or we can strip hydrogen from methane (cheap, efficient, but produces CO_2). Either way, hydrogen depends on external energy.

This is why calling hydrogen a "fuel" is as misleading as calling a battery a "fuel." Neither stores energy of its own.

Both must be charged.

And charging takes energy — real energy — extracted from the Earth in ways that are anything but green.

Hydrogen's difficulty does not stop there. It is the smallest molecule known, able to escape through microscopic imperfections in steel, welds, tanks, and valves. It embrittles metals, leaks easily, ignites at low energies, and requires extreme pressures or cryogenic temperatures to store in a meaningful way. Hydrogen is the least energy-dense fuel by volume that has ever been proposed for widespread use.

To move a car a meaningful distance, a hydrogen tank must either be enormous, thick-walled, refrigerated, or pressurized to extraordinary levels. The energy cost of creating, compressing, transporting, and storing hydrogen often exceeds the energy delivered when it is used.

This is not a path to a clean future.

This is a Rube Goldberg machine.

Hydrogen is not the next great fuel of civilization.

It is, in many cases, a wish.

The Battery Illusion

The same confusion surrounds batteries.

To the casual observer, a battery seems like a generator -- something that creates electricity. This illusion is reinforced by the experience of driving an electric vehicle or seeing a phone charge from a cable that looks passive. But a battery is merely an electrochemical reservoir, nothing more.

Tesla drivers are not driving on "stored electrons from the sun."

They are driving on the energy mix of their grid — almost always dominated by natural gas, coal, oil, nuclear, or hydroelectricity.

Every mile driven in an electric car is powered by the present-day reality of the grid, not by the battery itself.

Batteries are marvelous devices — elegant in their chemistry, efficient in their discharge, beautifully engineered — but they are not sources of energy. They do not generate power. They cannot lift civilization out of scarcity. They merely store.

And to store, they must be charged.

And to charge, they must have a source.

In the real world, that source is overwhelmingly fossil fuels — not because of conspiracy or negligence, but because fossil fuels remain the backbone of global electricity generation. Without natural gas, coal, and nuclear, there would be no electricity to charge a battery at all.

A battery-powered world is still a hydrocarbon-powered world, unless the entire grid behind the battery is transformed. This is why battery-only solutions cannot save the world. They merely shift the location of combustion from under the hood to behind the power plant fence.

Useful? Immensely.

Life-saving? No.

A civilization-scale solution? Not even close.

The Case of Hybrids

Hybrids occupy the narrow bridge between combustion and electrification. They are practical, efficient, and valuable tools. They save fuel. They reduce emissions under specific conditions. They integrate regenerative braking. They allow smaller engines to do more work.

But they do not break the energy equation.

A hybrid is still powered by gasoline.

Its battery is still powered by the grid.

It is a technological refinement, not a transformation.

Hybrids are clever, virtuous, reasonable — and utterly insufficient as a planetary solution. They are a stopgap. A chapter. A step.

Civilization does not run on steps.

Civilization runs on energy.

The Real Lesson of Hydrogen and Batteries

Hydrogen and batteries share a single truth:

They are not fuels.

They are storage devices.

Their usefulness depends on the energy we generate elsewhere.

And the energy we generate elsewhere still comes from:

- Natural gas
- Coal
- Oil
- Hydropower
- Nuclear fission

Solar and wind (themselves deeply dependent on hydrocarbons for materials, mining, manufacturing, installation, transport, and grid stabilization)

There is no path to a clean future that bypasses generation.

People misunderstand hydrogen and batteries because they misunderstand energy. They imagine that technology can replace physics, that cleverness can replace chemistry, that marketing can replace entropy.

But the truth is simple:

Civilization will not be saved by storing energy differently.

Civilization will be saved by producing energy differently.

And the only scalable, concentrated, non-intermittent, resource-abundant energy sources capable of replacing fossil fuels at a civilizational scale are found not in the winds or tides or biofuels, but in the atomic heart of matter itself.

Fission.

And eventually, fusion.

These are not alternatives.

They are the final trajectory of civilization's energy evolution.

Hydrogen and batteries are tools.

But fission and fusion are destiny.

CHAPTER 16
FISSION: BOTTLED E=MC²

How the unlocking of the atomic nucleus transformed our energy destiny and marked the dawn of humanity's cosmic adulthood.

The story of energy takes a sudden, shocking turn in the early 20th century — a turn so dramatic that it stands apart from every discovery before it. Coal, oil, and natural gas gave us power. Electricity gave us speed. But fission gave us something altogether new:

The ability to command the binding energy of atoms.

For the first time since stars ignited in the early universe, a species on a small rocky planet learned how to tap the fire that holds nuclei together.

This did not arise from a desire for electricity.

It emerged from physics — from an attempt to understand

the strange, mathematical world inside the atom. A world where energies defy intuition, where forces bend space, where particles behave as both waves and points, where matter exists in states evolution never prepared the human mind to grasp.

The path to fission began not in an engineering lab but in the introspection of thinkers like Marie Curie, Ernest Rutherford, Enrico Fermi, Otto Hahn, and Lise Meitner — people who stared into the invisible and saw structure. They uncovered a truth as old as the universe itself:

Atoms contain more energy than chemistry could ever dream of.

In 1938, Hahn and Strassmann split the uranium nucleus, and Meitner — in one of the most elegant acts of scientific interpretation ever written — realized what had happened. The nucleus had broken. Mass had vanished. Energy had appeared in its place.

$$E = mc^2$$

A quiet equation suddenly roared to life.

A pea-sized amount of uranium held more energy than tons of coal.

A handful of fissile atoms could power cities or destroy them.

Humanity had crossed a threshold with no precedent.

The Power Hidden in the Nucleus

Fission is not combustion; it is not oxidation; it is not chemistry of any kind. It is the release of binding energy — the very glue that holds protons and neutrons together. This energy sits locked beneath layers of electromagnetic forces, released only when specific atomic arrangements become unstable and crack apart.

Where hydrocarbons release tens of millions of joules per kilogram, uranium releases billions.

Per kilogram, nuclear fuel contains three million times more energy than oil.

Nothing in biology or geology prepared us for this magnitude.

This was not an evolutionary inheritance.

This was something new under the Sun.

Fission was the moment humanity stopped being bound to the surface of its own planet — energetically, intellectually, technologically, and morally.

The Tragic Detour

It is impossible to tell the story of fission without acknowledging its darkest chapter. The first great public demonstration of nuclear power was not a power plant but a weapon. The Manhattan Project accelerated the physics, but it also carved a wound into the global psyche — a wound that still shapes public perception of nuclear energy.

Fear overshadowed understanding.

Emotion overshadowed physics.

Politics overshadowed the thermodynamic truth.

The tragedy is not that we built weapons.

The tragedy is that we stopped building power plants.

We turned away from a technology that could have given us clean, abundant energy for centuries. We let fear freeze progress. We let rhetoric overshadow reason. We let myths derail a civilizational breakthrough.

Yet fission remains as true and as necessary now as it was in 1945.

The Reality: Fission Is the Cleanest, Densest, Most Scalable Energy Ever Deployed

Nuclear fission produces:

- virtually no CO_2
- no soot
- no methane
- no sulfur
- no particulates
- no reliance on vast landscapes or weather
- no intermittency
- no fragile supply chains
- no need for backup fossil plants

And its fuel — uranium — is astonishingly abundant. A single fuel pellet the size of your fingertip contains the energy of an entire ton of coal.

Reactors operate for 18 to 24 months without refueling.

Their waste volume is tiny.

Their safety record, despite cultural fear, is among the best of any major technology ever invented.

Chernobyl was a unique Soviet design without containment.

Fukushima was a 1960s-era plant hit with a once-in-a-millennium tsunami.

Neither says anything about the reactors we build today.

Modern plants are sealed fortresses of physics and engineering.

Small Modular Reactors (SMRs) are even safer — walk-away safe, passively cooled, meltdown-proof. They will scale before fusion arrives.

If humanity were rational, we would be building dozens every year.

The First Step Toward the Stars

A species that relies on photosynthesis and hydrocarbons is a surface species.

A species that harnesses fission becomes a cosmic species.

Because fission is the first energy source not tied to the biosphere.

It lifts us off the constraints of weather, sunlight, forests, land area, and real-time energy production. It gives us a power dense enough to launch ships into deep space, power bases on the Moon, melt ice on Europa, drill into Titan, and one day propel starships at fractions of light-speed.

Coal let us break free from muscle.

Oil let us break free from distance.

Gas let us bridge the present.

However, fission is the first technology that enables us to break free from Earth.

It is our first cosmic engine.

And it is the necessary stepping-stone to fusion — the crowned jewel of energy, the destiny of our species.

Humanity's moral duty is not just to survive.

It is to perpetuate consciousness in a universe that produces very little of it.

And that requires energy not of wind or water, but of nuclei.

Fission was the first human-made starfire.

It was the moment we stepped beyond our origins and caught a glimpse of what we could become.

Addendum to Chapter 16 — The Power, Safety, Promise, and Lost Potential of Fission

How old reactor designs shaped fear, how politics derailed progress, and how modern nuclear technology is safer, cleaner, and more necessary than ever.

Nuclear fission is not a single technology.

It is a family of technologies — some obsolete, some brilliant, some never built, some still waiting for their moment. The tragedy of the nuclear age is simple: we built the wrong reactors first, and then we judged the entire field by those early missteps.

The world's first commercial reactors were not designed for safety, efficiency, or electricity production. They were designed for one reason:

To produce plutonium for weapons.

This is the original sin of nuclear energy — not that it is dangerous, but that its infancy was shaped by geopolitics instead of engineering purity.

The Early Reactors: Safe Enough, But Not What They Could Have Been

The pressurized water reactor (PWR) and boiling water reactor (BWR) — the workhorses of the mid-20th century — were robust designs, but they were not optimized for long-term civilian energy production. They used uranium-235, a scarce isotope. They operated at high pressures. They required enormous containment buildings. They produced long-lived waste.

Yet even so, their safety record was extraordinary.

Three Mile Island harmed no one.

Fukushima killed no one from radiation.

Western reactors have among the lowest mortality rates of any energy technology ever deployed.

Despite decades of fear, nuclear energy has caused vastly fewer deaths than coal, oil, natural gas, biomass, or even the manufacturing supply chains of solar energy. And yet, the public imagination turned against it.

Not because of nuclear power.

But because of nuclear weapons.

Chernobyl, often cited, was not a nuclear plant in the modern sense. It had no containment dome. It used a Soviet graphite-moderated design that would never be licensed today. It was the product of a geopolitical system that valued secrecy over safety.

The problem was not nuclear energy.

The problem was the Cold War.

The Missed Miracle: Thorium

If the story ended there, it would already be a tragedy.

But what makes it heartbreaking is this:

We had a better solution in hand.

We simply chose not to pursue it.

Thorium — element 90 on the periodic table — is one of the most abundant actinides on Earth.

It is fertile (meaning it can become fissile with neutron exposure), but not easily weaponized.

It produces far less long-lived waste.

It cannot melt down in the catastrophic sense.

And it burns almost all of its fuel, leaving orders of magnitude less material to store.

In the 1960s, the Oak Ridge National Laboratory constructed the Molten Salt Reactor Experiment (MSRE) — a thorium-capable design utilizing liquid fuel. It ran beautifully. It proved the physics. It solved the meltdown problem by eliminating the solid fuel rods altogether. A molten-salt reactor simply drains its fuel into a safe, subcritical tank if power is lost.

It was elegant.

It was safe.

It was the future.

And it was killed.

Why?

Because a thorium fuel cycle does not easily produce plutonium for bombs, uranium did. Thorium did not. And in the geopolitical logic of the Cold War, weapons pathways were prioritized over energy pathways.

Thus, the greatest opportunity for clean energy in human history was shelved for political convenience.

Small Modular Reactors: What We Should Have Built All Along

Today, the reactors we are finally beginning to build look much more like what we should have built 60 years ago.

Small Modular Reactors (SMRs) are:

- factory-built
- transportable
- walk-away safe
- passively cooled
- meltdown-proof by design
- low-pressure
- designed to shut themselves down with no human intervention

Many use molten salt, high-temperature gas-cooled designs, or fast-spectrum neutronics that can burn fuel once considered "waste."

They can be buried in the ground.

They can run for 10–20 years without refueling.

They can provide baseload power without emissions, smoke, soot, or dependence on weather.

They will power mining operations, villages, factories, desalination plants, data centers, and space missions.

This is the nuclear age we deserved the first time.

This is the nuclear age we can still build.

The Geopolitical and Nonproliferation Reality

A truth rarely spoken:

Nuclear energy is stabilizing, not destabilizing.

Nations with reliable fission-based electricity have:

- greater energy security
- less dependence on fuel imports
- more stable economies
- fewer incentives for conflict
- fewer reasons for geopolitical leverage over pipelines and straits

And contrary to public fear, the proliferation risk from modern reactors is negligible. Many SMRs are designed specifically to eliminate pathways for weapons. Fuel remains sealed, unextractable, unseparable. The United States, Canada, Japan, South Korea, France, and the UK are all developing reactors with no plutonium production pathways.

Thorium cycles, molten salt designs, and high-assay, low-enriched uranium (HALEU) systems all significantly mitigate proliferation risks.

The world needs energy.

Nuclear energy provides it safely.

And it reduces the geopolitical tension created by fossil fuel scarcity and distribution.

The Central Truth: Nuclear Energy Is the Only Scalable Clean Energy We Already Have

Wind and solar are useful. Batteries are useful. Hydrogen has niche applications. But none of them can replace fossil fuels a global scale. They lack the power density, reliability, and material abundance. They cannot power steel, cement, aviation, or industrial heat. They cannot run 24/7. They cannot stand alone.

Nuclear can.

And nuclear must.

If humanity is serious about decarbonization — without plunging civilization into scarcity — there is only one answer available now: build reactors.

Not someday.

Not after endless debate.

Now.

A Call to Action: What the Reader Should Feel

If this addendum has worked — and I believe it has — then the reader's emotional response should be a blend of exhilaration and indignation:

"Why weren't we told this?

Why did we waste decades?

Why aren't we building reactors everywhere?"

Good.

That is exactly what you wanted.

Because this chapter isn't just about fission.

It is about the moral obligation of a species on the threshold of the stars.

We must choose abundance over fear.

We must choose engineering over nostalgia.

We must choose reality over rhetoric.

We must choose the future over the past.

They are not theoretical.

They are not "on the horizon."

They already exist — at scale, in the real world, in the most demanding environments humanity operates.

Additional Section for the Addendum — The Proven Power of Modular Reactors

There is a persistent misconception that small modular reactors (SMRs) are untested innovations — futuristic ideas waiting for validation. The opposite is true. Humanity has been running modular nuclear reactors for over 60 years in the most hostile, mission-critical environments imaginable.

The U.S. Navy has operated hundreds of compact reactors in submarines and aircraft carriers, each one functioning as a sealed, self-contained, walk-away-safe power plant. These reactors endure decades of vibration, pressure fluctuations, violent maneuvers, shock events, and operational extremes that no civilian power station will ever face.

And yet, the Navy's nuclear propulsion record is pristine.

No meltdowns.

No mass casualties.

No radiological disasters.

Just millions of nautical miles of clean, continuous, reliable power.

If you wanted a proof-of-concept for mass-produced, robust, small-scale fission reactors… You already have it.

These naval reactors are:

- modular
- refuelable only after years of operation

- designed to run quietly and safely under enormous stress
- maintained by crews with training but not PhDs in nuclear physics

Their reliability is not a theoretical claim.

It is a seven-decade empirical fact.

This experience reveals something profound:

We know how to build small reactors.

We know how to mass-produce them.

And we know how to operate them safely.

The only thing missing is the will to deploy them on land.

SMRs for Communities, Industry, and Critical Infrastructure

The brilliance of civilian SMR designs lies in their flexibility. They can be produced in factories, shipped by rail or truck, and installed on-site like industrial equipment. This changes everything.

Imagine:

- new housing subdivisions powered by a 50–100 MW SMR
- metropolitan hospitals with their own on-site, meltdown-proof reactor guaranteeing uninterrupted power during disasters

- small towns liberated from fragile, aging grids
- data centers running on nuclear reliability instead of diesel backups
- factories, desalination plants, and mines powered by constant, carbon-free heat
- remote regions freed from diesel fuel deliveries
- emergency shelters powered indefinitely during crises
- military installations using SMRs for hardened, secure power
- island nations replacing bunker fuel with safe nuclear heat

These are not fantasies.

These are engineering blueprints with regulatory footprints already being developed across the U.S., Canada, France, Korea, and Japan.

A 21st-century SMR behaves like a "nuclear appliance" — compact, sealed, requiring minimal maintenance, and designed for passive safety. If power is lost, cooling continues without the need for pumps, pipes, operators, or electricity. Heat dissipates naturally. The reactor's physics keep it safe.

You cannot melt down what is engineered not to melt.

The Maintenance Myth

A common fear is that nuclear plants require vast, specialized teams. SMRs overturn that assumption entirely. They are:

- sealed for their entire operational life
- designed for outsourced, periodic servicing
- automated to a degree unthinkable in early reactors
- monitored 24/7 from centralized control centers
- For communities, this means a subdivision or small town could have a nuclear plant that:
- requires no on-site specialists
- operates for a decade or more without refueling
- fails safe by design, not by intervention
- provides stable electricity without the fragility of renewables or long-distance transmission

It is the opposite of dangerous.

It is the definition of secure.

The Real Question: If This Works at Sea, Why Not on Land?

It is a strangely overlooked truth:

An aircraft carrier is essentially a floating city powered entirely by nuclear energy.

Thousands of people.

Hospitals.

Radar systems.

Desalination plants.

Weapons systems.

Computers.

Lighting.

Heating and cooling.

Fresh water and life-support.

All powered quietly, cleanly, continuously by a pair of shipboard reactors.

If we trust nuclear power for a vessel carrying 5,000 lives and billions of dollars of equipment —

if we entrust it to sailors in their twenties —

if we operate it under wartime conditions —

then surely we can trust a modern SMR sitting under a concrete dome in Nevada, Kansas, Ontario, or Tokyo.

This is not an energy risk.

This is an energy revolution waiting to be embraced.

The technology exists.

The prototypes are built.

The safety is proven.

The need is urgent.

The only barrier is fear — the residue of a bygone geopolitical age.

CHAPTER 17

FUSION: LIGHTING A STAR ON EARTH

Tokamaks, stellarators, lasers, magnets, materials, AI — and why fusion is inevitable, but not imminent.

To grasp the meaning of fusion, one must first step out of the smallness of human history and into the furnace of the cosmos. Every star we see — every point of light in the night sky, every glowing nebula, every sun in every galaxy — is powered by fusion. It is the first great energy source of the universe, the one that ignites hydrogen into helium, light into heat, gravity into radiance.

We are star-born beings living on a small rocky outpost forged from the ashes of ancient suns. Our entire existence rests on the quiet arithmetic of fusion reactions occurring 93 million miles away, in the Sun's core, where pressure and temperature weave protons into heavier nuclei.

But on Earth, we do not live inside a gravitational well

capable of crushing hydrogen into union. We live at atmospheric pressure, surrounded by cool oceans and temperate skies. For most of human history, fusion was a distant miracle, a cosmic phenomenon beyond imitation.

And yet — for the first time in planetary history — a species has begun to learn how to make its own sunfire.

The Promise of Fusion

Fusion is conceptually simple:

Join light nuclei (usually isotopes of hydrogen).

Overcome their electric repulsion.

Let the strong nuclear force bind them.

Release vast amounts of energy.

It is the mirror of fission — not breaking heavy atoms but merging small ones. And the energy released per reaction is extraordinary.

A few grams of deuterium and tritium can produce as much energy as tons of coal, millions of AA batteries, or barrels of oil. Fusion releases no carbon, produces no long-lived waste, and cannot melt down. It is, in every respect, the ideal energy source for an industrial civilization — dense, clean, inexhaustible.

So why don't we have it yet?

Because nature cheats.

Stars use gravity.

On Earth, we must use physics.

Tokamaks: The Magnetic Bottle

The most mature fusion technology is the tokamak, invented in the Soviet Union. It traps plasma — gas heated so hot that electrons and nuclei separate — using magnetic fields wrapped in a torus. The plasma must be:

- so hot (100 million degrees)
- so stable (no turbulence)
- so dense
- and so well-confined
- that fusion outpaces the heat losses. This is the Lawson Criterion, the three-part test for ignition.

Tokamaks have come astoundingly close. JET in the UK, EAST in China, KSTAR in Korea, and many others have all achieved record temperatures, confinement times, and fusion power output. ITER, the international megaproject, aims to demonstrate tenfold energy gain.

But tokamaks are devilishly difficult. Plasma behaves like weather in a tornado of magnetic chaos. It writhes, buckles, oscillates, and must be controlled with exquisite precision. This is where modern AI comes in: unprecedented machine-learning algorithms are beginning to stabilize plasmas in ways that human operators never could.

Tokamaks are not a fantasy.

They are an engineering challenge.

And they are within reach.

Stellarators: The Baroque Beauty

If tokamaks are elegant in concept but difficult in practice, stellarators are the opposite: wildly complex machines that solve stability at the cost of geometry. Their coils twist like the backbone of a mythical creature, stabilizing plasma naturally without the same need for induced current.

The Wendelstein 7-X stellarator in Germany is a monument to computational design — a machine whose electromagnetic coils look like sculpture but behave like mathematical perfection.

Stellarators are stable.

Tokamaks are powerful.

Both are paths to the same star.

Inertial Confinement: The Laser Star

Then there is inertial confinement fusion — a path entirely different from the others. Instead of using magnetic fields to hold plasma steady, inertial confinement uses brute-force physics: lasers or particle beams aimed at a small fuel pellet, compressing it so violently and so rapidly that fusion occurs before the pellet has time to expand.

The National Ignition Facility achieved ignition in 2022 — a historic milestone, demonstrating for the first time on Earth a controlled fusion reaction that produced more energy than the fuel absorbed.

But inertial confinement is not ready for power plants.

Its lasers are massive, its fuel impractical, and its repetition rate too low.

Yet the breakthrough is real.

Fusion is not a dream.

Fusion is a frontier.

The Tritium Bottleneck

Here is the most important part the public does not understand:

Fusion is not limited by physics.

It is limited by tritium.

Deuterium is abundant — it can be extracted from seawater. But tritium, the other half of the deuterium–tritium reaction, is rare, radioactive, and decays with a 12.3-year half-life. The world's tritium supply — produced mainly in Canadian CANDU reactors — is measured in kilograms.

A commercial-scale fusion industry will require more tritium than exists.

This is why modern fusion reactors must be designed with breeding blankets — layers of lithium that absorb neutrons and generate new tritium via nuclear reactions. This is solvable. It just requires engineering. But it is the most significant bottleneck in fusion deployment over the next decades.

Why Fusion Is Inevitable, But Not Imminent?

Fusion is not a matter of "if."

It is a matter of "when."

Humanity will harness fusion because the physics is sound. The energy is too abundant to ignore. The demand too great. The stakes too high. Fusion is the logical endpoint of civilization's energy evolution.

But it will not arrive tomorrow.

We must create:

- new materials that can withstand neutron bombardment
- new magnets — superconducting, cryogenic, AI-controlled
- new tritium breeding cycles
- new plasma models
- new designs for heat extraction
- new regulatory frameworks
- new supply chains

This will take decades.

But decades are not centuries.

And decades are not never.

Fusion is inevitable.

Because the laws of physics demand it.

And because humanity — if it intends to survive long enough to become a spacefaring species — has no other choice.

Fusion and the Human Imperative

Fusion is more than technology.

More than science.

More than engineering.

It is the fulfillment of the arc that began when the first microbe split water for energy, when the first forest fell under human fire, when the first steam engine shook the English countryside, when the first uranium nucleus cracked apart, and when the first human mind looked at the night sky and felt the pull of the stars.

Fusion is the moment when Earth stops being a prison of scarcity and becomes a cradle of abundance.

It is the moment when human civilization gains the power to:

- terraform
- explore

- colonize
- protect
- endure
- and flourish
- without consuming the biosphere that created it.

The energy of the stars will be the energy of the Earth.

And when that happens, the universe will no longer be silent — not because someone else speaks, but because we finally will.

This is the great turning in the story of life.

CHAPTER 18

THE GREAT SILENCE REVISITED

Why energy scarcity — not self-destruction — is the real answer to the Fermi Paradox. Civilization as a thermodynamic phenomenon.

For decades, the Fermi Paradox has haunted the scientific imagination. Enrico Fermi captured it in three devastating words:

"Where is everybody?"

The universe is old.

The universe is vast.

The universe is fertile with stars, planets, and chemistry.

Life emerged early on Earth — almost immediately once the crust cooled and oceans formed. The cosmic ingredients for biology are abundant: carbon, hydrogen, nitrogen, oxygen, phosphorus, and sulfur. Amino acids form in comets. Organic molecules drift between stars.

Life is not the mystery.

Life is the default.

The real mystery is civilization.

Where are the radio signals?

Where are the artifacts?

Where are the megastructures?

Where are the probes?

Where are the footprints of intelligence in a galaxy that should have had billions of chances?

The silence is not just puzzling.

It is oppressive.

It is cosmic.

But perhaps the problem lies not in our pessimism, nor in extraterrestrial fragility, nor in cosmic violence, nor in self-destruction, nor in the hubris of expecting company.

Perhaps the answer is simpler, more elegant, more devastating:

Civilization is a high-energy phenomenon.

And the universe almost never provides the fuel.

The Thermodynamic Threshold

Civilization is not simply intelligence.

Civilization is the sustained ability of a species to concentrate, store, transform, and deploy energy at scales far beyond its biological baseline. Ants build societies. Octopuses solve puzzles. Crows make tools. Dolphins communicate. But none of these species crossed the energetic threshold required to rewrite their environment, amplify their productivity, or create technologies that compound over generations.

Civilization begins when energy abundance begins.

And energy abundance is rare.

On Earth, the rise of technological civilization did not depend solely on intelligence. It depended on:

- a massive reservoir of buried carbon
- an improbable history of atmospheric oxygenation
- a uniquely large moon stabilizing climate
- prolonged continental stability
- plate tectonics recycling nutrients
- and hundreds of millions of years of uninterrupted photosynthesis to store energy

This was not merely biological luck.

It was energetic fortune.

Civilization is not a cultural artifact.

It is a thermodynamic feedback loop.

Break the energy, and you break the loop.

The Universe of Almost-Civilizations

If microbial life is common — and all evidence suggests it is — then planets teeming with algae, bacteria, and simple multicellular organisms may be everywhere. But a world with microbes is not a world with metallurgy.

Energy scarcity suffocates advancement long before technology can begin.

Consider the requirements:

1. Metallurgy requires fire hotter than wood can provide.
2. Industry requires storable, dense energy.
3. Electricity requires generators and turbines.
4. Electronically mediated civilization requires massive surplus power.
5. Spaceflight requires chemical, nuclear, or fusion fuels.

Without fossil fuels or their equivalents, no world reaches industrialization.

Without nuclear fission, no world sustains it.

Without fusion, no world spreads beyond its star.

What fraction of habitable planets possess:

- large, accessible coal seams?
- petroleum reserves formed by ancient oceans?
- stable continents with erosion exposing those seams?
- enough atmospheric oxygen for combustion?
- the right minerals for metallurgy and electronics?
- uranium and thorium enriched by a nearby supernova?

We already know the answer.

Almost none.

Earth is not just a habitable world.

Earth is a high-energy world.

And high-energy worlds are cosmically exceptional.

Why We Do Not Hear Them

Civilizations do not need to destroy themselves.

Most simply never ignite.

They may develop intelligence — but without energy density, they plateau. Their technology remains artisanal, not industrial. Their communication remains local, not electromagnetic. Their mobility remains biological, not mechanical. Their structures remain organic, not metallic.

The universe may be full of minds.

But minds without megawatts remain silent.

A species can be brilliant and never leave its caves.

Brilliant and never refine metal.

Brilliant and never build an engine.

Brilliant and never develop radio waves.

Brilliant and never escape its biosphere.

The universe is not empty.

It is incomplete.

Intelligence is not enough.

Progress is not enough.

Curiosity is not enough.

Energy is the limiting reagent of civilization.

Remove fossil fuels from Earth's crust, and humanity would never have escaped the photosynthetic trap.

Remove nuclear isotopes from Earth's mantle, and we would never escape fossil fuels.

Remove the possibility of fusion, and we will never escape Earth.

Civilizations do not die.

Civilizations fail to launch.

The Great Silence Is a Thermodynamic Silence

The Fermi Paradox dissolves once civilization is understood as an energy-dependent phenomenon.

The silence is not ominous.

It is not tragic.

It is not the echo of apocalypse.

It is simply the default state of a universe where almost no planet acquires the geological inheritance necessary to lift life from biology into technology.

And Earth — improbable Earth — is that inheritance.

Our silence is not a warning.

It is a privilege.

We are not alone because we are solitary.

We are alone because we are energetically rare.

Addendum: Other Answers to the Great Silence

Over the decades, many explanations for the Fermi Paradox have been proposed. Most are thoughtful. Many are internally consistent. None are frivolous. But nearly all share a common assumption: that civilization is easy to ignite, yet fragile to sustain.

My work challenges that assumption.

Below, this book briefly acknowledges the most commonly cited explanations for the Great Silence — not to dismiss them, but to place them in thermodynamic context.

1. Self-Destruction

Perhaps civilizations arise frequently, but destroy themselves through war, environmental collapse, artificial intelligence, or technological excess.

This hypothesis assumes a universe teeming with industrial civilizations that routinely reach advanced technological stages — only to erase themselves shortly thereafter.

My position differs:

Most civilizations never reach the energy abundance required to become dangerous in the first place. One cannot destroy oneself with nuclear weapons one never had the surplus energy to build.

Self-destruction may occur — but it cannot explain the silence alone.

2. The Great Filter

The Great Filter proposes that one or more steps between simple life and interstellar civilization are extraordinarily unlikely — and that humanity has either already passed it, or is yet to face it.

The Adair Hypothesis reframes the filter not as a mysterious biological or sociological hurdle, but as a geological and energetic one.

The filter is not intelligence.

The filter is not cooperation.

The filter is not morality.

The filter is energy density.

3. Rare Earth

The Rare Earth hypothesis argues that complex life depends on an unlikely confluence of planetary conditions — stabilizing moons, plate tectonics, magnetic fields, and long-term climate stability.

I agree — but extend it further.

A planet can host complex life and still lack the energetic inheritance required for civilization. Earth is not merely rare because it hosts life; it is rare because it stockpiled energy over geological time in forms accessible to intelligence.

Habitability is not enough.

4. They Are Out There, But Quiet

Some suggest advanced civilizations deliberately avoid detection, communicate in unknown ways, or have moved beyond radio signals entirely.

This explanation presupposes civilizations that achieved immense technological maturity — quietly and ubiquitously.

I suggest a simpler alternative:

Silence does not require concealment if ignition never occurred.

5. We Are Early

Another possibility is that humanity is among the first technological civilizations to arise — that the universe simply hasn't had time.

This may be partially true. But time alone does not create coal seams, fissile isotopes, or fusion pathways. Geological fortune does not scale with cosmic age.

All of these explanations may operate at the margins. Some may be true in isolated cases. None need be rejected.

But none explain the *default* silence.

The Adair Hypothesis proposes that technological civilization is not the natural continuation of life, but a rare thermodynamic phase transition — one that requires extraordinary energy concentration unlikely to occur on most worlds.

The universe may be alive.

It may be intelligent.

It may even be curious.

But without dense, scalable energy, it remains quiet.

The Great Silence is not the absence of minds.

It is the absence of fuel.

This is the foundation of the Adair Hypothesis.

In the next chapter, we will bring it into full view.

CHAPTER 19

THE ADAIR HYPOTHESIS (FULL EXPOSITION)

Why is microbial life common, intellect possible, but technological civilization requires an almost unimaginable geological inheritance?

The universe is not empty.

It is indifferent.

It does not hide civilizations.

It fails to produce them.

For all its galaxies, nebulae, molecular clouds, star nurseries, and habitable-zone planets, the cosmos is profoundly poor at generating the one thing humans crave to find:

A technological species capable of signaling across the stars.

Life?

Life is everywhere.

Life blooms whenever chemistry and time embrace.

But civilization — true civilization — is another matter entirely. Civilization is not merely the presence of intelligence. Dolphins are intelligent. Cephalopods are intelligent. Corvids, elephants, apes, and even some fish demonstrate cognition, memory, emotion, planning. Intelligence is not the rarity.

Energy is the rarity.

And this is the central claim of the Adair Hypothesis:

Technological civilization is the product not of biology,

but of geology — and only planets with a specific, extraordinarily rare energy inheritance can ever cross the threshold.

Life Is Common

Organic molecules form in interstellar clouds.

Amino acids have been found in meteorites.

Complex hydrocarbons form on Titan.

Water worlds orbit red dwarfs in the trillions.

Wherever liquid water persists, microbial life appears to be the natural outcome. On Earth, life emerged almost as soon as oceans cooled — within a few hundred million years. This is fast on cosmic timescales.

Life does not require miracles.

Life requires chemistry.

But chemistry alone is not enough.

Intelligence Is Possible

Evolution is a tinkerer, not a designer. Given billions of years, it repeatedly stumbles toward complexity. Eyes have evolved independently multiple times. Flight has evolved multiple times. Camouflage, cooperation, tool use, communication, navigation — all have arisen across unrelated lineages.

Intelligence, too, appears likely to evolve when ecological pressures reward flexibility, memory, and planning. Many planets may host clever creatures capable of solving puzzles, building nests, even using simple tools.

But tool use is not technology.

Technology is not intelligence.

Technology is amplified energy.

And this is where almost all worlds fail.

Technological Civilization Requires Geological Fortune

Earth did not produce civilization because humans were clever.

Earth produced civilization because Earth was energetically gifted beyond all reason.

Our planet inherited:

1. Stable continents created by plate tectonics
2. An oxygen-rich atmosphere driven by a rare photosynthetic revolution
3. A stable climate moderated by a large moon
4. Enormous, accessible reserves of coal
5. Vast petroleum basins formed by ancient anoxic seas
6. Natural gas fields
7. High-grade uranium and thorium
8. A 4.5-billion-year accumulation of geological processes

This combination is absurdly unlikely.

Take any one away — coal, oil, oxygen, tectonics, radioisotopes — and Earth remains a beautiful biosphere but never becomes a technological civilization.

Imagine a world:

- without fossil fuels (98% of planets)
- without accessible uranium (most planets)
- without atmospheric oxygen (almost all planets)
- without long-term carbon burial (most planets)
- without plate tectonics to cycle nutrients (unknown fraction, likely large)
- with oceans too deep to expose continents
- with a thick shroud of clouds blocking starlight
- with tidal locking around a red dwarf

- with frequent sterilizing flares

Life could flourish.

Brains could evolve.

Intelligence could emerge.

But without dense, storable, scalable energy, no world reaches electricity, computing, metallurgy, or propulsion.

Intelligence might dream.

But it cannot build.

Civilization Is a Thermodynamic Feedback Loop

This is the heart of the hypothesis:

Civilization is not a cultural event.

It is a thermodynamic transition.

Civilization begins when a species learns to harness energy beyond biology:

- Fire
- Metallurgy
- Steam
- Fossil fuels
- Electricity
- Fission
- (Eventually) fusion

Each step increases the energy throughput of the species, allowing exponential growth in complexity.

But that first ignition — the escape from the photosynthetic trap — is possible only on worlds with a deep-time energy inheritance.

Remove fossil fuels from Earth's crust, and humans never leave the Iron Age.

Remove uranium, and we stagnate after fossil depletion.

Remove fusion potential, and no civilization ever spreads among the stars.

This is not pessimism.

This is physics.

Most Planets Are Not Energetic Enough

The galaxies may be filled with life-bearing worlds, but they are:

1. geophysically stagnant
2. energetically poor
3. geomineralogically limited
4. lacking in buried carbon
5. lacking in accessible radioisotopes
6. lacking in oxygen
7. lacking in stable climates
8. lacking in the conditions for industrial ignition

On such planets, evolution may produce minds, but not machines.

The universe is not teeming with starships.

It is teeming with almost-civilizations — worlds capable of thought, incapable of progress.

This is the Great Silence.

Not a graveyard.

Not a cosmic genocide.

Not self-destruction.

Simply:

energetic poverty.

Civilization is hard.

Energy is scarce.

Geological fortune is rare.

Earth is one of the luckiest planets in the galaxy.

The Adair Hypothesis, Stated Formally

Microbial life is common.

Sentience is possible.

But technological civilization requires:

1. Large-scale carbon burial across hundreds of millions of years
2. Stable continents with exposed sedimentary basins
3. An oxygenated atmosphere capable of combustion
4. Abundant fossil hydrocarbons to escape the pre-industrial trap
5. Accessible uranium and thorium to escape fossil depletion
6. Fusion potential (a universal constant) for long-term interstellar survival

The simultaneous presence of all six is so astronomically rare that Earth may be the only known example within thousands or millions of light-years.

That is why the universe is silent.

Not because intelligence is rare.

But because energy inheritance is rare.

Not because civilizations destroy themselves.

But because almost none ever begin.

We are not witnesses to cosmic solitude.

We are witnesses to cosmic privilege.

And with privilege comes responsibility.

The next chapter makes that responsibility explicit.

CHAPTER 20

THE DUTY OF CIVILIZATION

Why energy abundance must transition from fossil fuels to fission and ultimately fusion.

Why failure to transition is not a moral failure, but scientific illiteracy.

Civilization is a fragile miracle sustained by something invisible yet omnipresent: energy. Every city skyline, every hospital, every semiconductor, every farm, every warm home, every vaccine refrigerator, every satellite, every electric light — all of it depends on our ability to control energy flows through matter.

This is not a political statement.

It is a physical one.

And as we have seen throughout this book, the story of

human civilization is the story of accessing increasingly powerful energy sources:

- Wood
- Coal
- Oil
- Natural gas
- Fission
- (Soon) fusion

Each stage lifted humanity to a new plateau of capability.

But with each step upward, the margin for misunderstanding grew.

We entered a world where energy is essential but invisible — where most people never witness a turbine, a reactor, a refinery, or a transmission line. Modernity hides its mechanisms. Comfort conceals the physics.

And so, a strange paradox emerged:

We depend on energy more than any previous generation, yet we understand energy less than any previous generation.

This is not a moral failure.

It is a gap between experience and infrastructure.

But bridging that gap — gently, clearly, welcomingly — is the duty of civilization.

Fossil Fuels: Our First Escape from the Trap

We owe everything to fossil fuels.

Everything modern civilization has achieved was made possible by the buried sunlight of ancient oceans and forests. They lifted us out of the biomass trap, multiplied human labor thousands of times, and fueled agriculture, medicine, transportation, and industry.

Coal powered steam.

Oil powered engines.

Natural gas powered electricity with unmatched reliability.

They were not villains.

They were our scaffolding.

But scaffolding is not meant to be permanent.

Fossil fuels come with limits — geological, environmental, and economic. They are finite. They are unevenly distributed. And they increasingly strain the planetary systems that sustain us.

Acknowledging this is not condemnation.

It is maturity.

A civilization must know when to move on.

The Transition to Fission: The Adult Energy of an Industrial World

Nuclear fission is not a relic of the Cold War.

It is the first truly post-scarcity energy source humanity ever discovered.

A single uranium fuel pellet — no larger than the tip of your finger — contains as much energy as:

- a ton of coal
- three barrels of oil
- 1,000 cubic feet of natural gas

No emissions.

No soot.

No smoke.

No carbon.

For sixty years, nuclear reactors have safely powered cities, aircraft carriers, submarines, medical isotope facilities, and research centers. They have saved millions of lives by displacing air pollution. They have produced more clean electricity than all wind and solar combined.

And yet, their promise remains unfulfilled.

Not because of physics.

Not because of safety.

But because of misunderstanding.

Fission is the technology of a civilization that wants to endure.

It is the bridge between the age of chemical energy and the age of stellar energy.

And it is a duty — not to the past, but to the future — to use it wisely, widely, and well.

Small Modular Reactors: The New Neighborhood Power Plant

The next generation of nuclear power is not the sprawling concrete megastructure of the 1970s.

It is compact, modular, and factory-built.

Small modular reactors (SMRs):

- can be mass-produced
- are passively safe
- require minimal maintenance
- are ideal for communities, hospitals, industry, and remote regions
- do not emit carbon
- do not depend on geography or weather
- operate continuously

They are the energy equivalent of the microchip: smaller, safer, more scalable, more reliable.

Deploying them is not radical.

It is responsible.

Fusion: The Endgame of Energy

Fusion is not a dream of futurists.

It is the natural endpoint of everything physics makes possible for a technological civilization.

Fusion is:

- inexhaustible
- carbon-free
- meltdown-proof
- cleaner than fission
- safer than fossil fuels
- scalable to any demand
- capable of powering spaceflight, terraforming, and planetary stewardship

Fusion is not imminent.

But it is inevitable.

And it is the duty of civilization to prepare for it, invest in it, and build the scientific literacy required to understand it.

Fusion is the moment when humanity ceases to be bound by the accidents of geography and the scarcity of geology — when civilization becomes a self-sustaining thermodynamic engine capable of lighting worlds.

“This duty is not merely moral — it is material.

It arises not from ideology, but from physics."

This chapter is not a scolding.

It is an invitation.

The duty of civilization is not:

- to feel shame
- to blame the past
- to punish the present
- to moralize energy
- to divide society into believers and deniers

The duty of civilization is:

- to understand the physics
- to grasp the inheritance
- to accept responsibility
- to build wisely
- to communicate clearly
- to teach generously
- to preserve what sustains us

There is no guilt in using fossil fuels.

There is only wisdom in transcending them.

There is no shame in fearing nuclear power.

There is only strength in learning what it truly is.

There is no arrogance in pursuing fusion.

There is only courage in reaching for abundance.

A civilization that understands its energy foundations becomes not only more powerful — it becomes more compassionate, more unified, and more capable of building a stable, peaceful, luminous world.

The Duty of Civilization Is the Duty to Continue

To understand energy is to understand continuity.

To understand continuity is to understand responsibility.

To understand responsibility is to understand the fabric of civilization itself.

We are the only species in the galaxy — perhaps the only one for many thousands of light-years — to have crossed the thermodynamic threshold from biology to technology. That inheritance does not call for guilt. It calls for stewardship.

A species that can light a star on Earth carries a weight no other species has ever borne.

And that brings us to the final chapter —

not a warning, but a benediction.

EPILOGUE: THE SILENT UNIVERSE AND THE LUMINOUS EARTH

A CALL TO STEWARDSHIP, REASON, HUMILITY, AND WONDER.

The universe is vast beyond comprehension — a cathedral of dark matter and light, stitched together by gravity and time. Its stars are uncountable. Its planets number in the trillions. Its galaxies drift like islands across an endless cosmic sea.

Yet for all its grandeur, the universe is quiet.

No signals.

No beacons.

No footsteps across the interstellar dark.

Just silence — ancient, steady, and complete.

For most of human history, that silence was terrifying. It implied loneliness, insignificance, or the specter of civilizations extinguished before they could speak.

But as we have journeyed through the deep logic of energy and life, the silence has taken on a different tone — not foreboding, but illuminating.

The silence is not a graveyard.

It is a revelation.

It tells us that Earth is not simply a planet.

Earth is a geological improbability.

A physical masterpiece.

A thermodynamic miracle wrapped in a thin shell of atmosphere and water.

We are not one among many.

We are one among almost none.

And that realization carries with it a profound responsibility.

The Weight of Inheritance

Every breath of oxygen we take was produced by microbial creatures more ancient than imagination.

Every calorie of food we eat was powered by starlight caught in the chemical cages of photosynthetic life.

Every watt of electricity that lights our cities, cools our hospitals, and powers our tools is the product of Earth's rare gift: a buried reservoir of ancient energy, amassed over eons, waiting for a species capable of unlocking it.

We were born into abundance we did not create.

We inherited a world more generous than probability would allow.

And we do not owe guilt for this inheritance.

We owe gratitude — and stewardship.

The Duty of the Luminous Planet

Humanity's story is not one of exploitation.

It is one of transcendence.

We broke the photosynthetic ceiling with coal.

We built a global civilization with oil.

We lit our cities with natural gas.

We split the atom to harness the power of the nucleus.

And someday — someday soon — we will ignite fusion and light a star on Earth.

This is not arrogance.

It is acceptance of our nature as a technological species.

The duty of the luminous planet is not to recoil from its power, but to understand it.

The duty is not to fear our tools, but to wield them wisely.

Not to undo civilization, but to refine it.

Not to retreat into nostalgia, but to advance into stewardship.

Civilization is not the enemy of nature.

Civilization is nature waking up.

The Silence as Permission

When a child wakes in a silent house, they whisper.

When a species wakes in a silent galaxy, it should do the same.

The silence around us is not rejection.

It is permission.

Permission to explore.

Permission to aspire.

Permission to learn.

Permission to build.

Permission to endure.

Permission to fill the silence with our own message — not of dominance, but of presence.

We are the voice the universe has not yet heard.

We are the message no other world could send.

We are the emissaries of a planet that beat the odds.

Civilization is not a burden.

It is a torch.

The Luminous Earth

Someday, perhaps millions of years from now, travelers from Earth — biological, synthetic, hybrid, or something beyond our imagination — will cross the light-years to other worlds. They will look back at the tiny blue dot that gave them everything.

They will know what we have begun to understand now:

that Earth is not ordinary

that Earth is not typical

that Earth is not expected

Earth is luminous.

A lone bright ember in a dark and ancient universe.

And they will know something else — something quiet and steady and profound:

We chose to survive.

We chose to learn.

We chose to understand the duty that comes with improbable existence.

The universe may be silent.

But Earth is not.

Not anymore.

We speak through science.

We speak through reason.

We speak through humility.

We speak through the careful stewardship of a rare world entrusted to us by deep time.

And perhaps, one day, we will speak across the stars.

FURTHER READING & SOURCES OF INSPIRATION

- **Carl Sagan — Cosmos:** The book that taught a generation how to think cosmically — with humility, rigor, and wonder.
- **Richard Dawkins — The Selfish Gene:** A foundational exploration of evolutionary logic and replicator-driven complexity.
- **Nick Lane — The Vital Question:** A brilliant examination of energy flow, mitochondria, and the biochemical origins of complexity.
- **Brian Greene — The Elegant Universe:** A sweeping tour through the structures of physical reality, from strings to spacetime.
- **Stephen Hawking — A Brief History of Time:** A poetic invitation into the deep architecture of the cosmos.

- **Jared Diamond — Guns, Germs, and Steel:** A masterwork on how geography and resources shaped human civilizations.
- **David Deutsch — The Beginning of Infinity:** A profound argument for optimism, progress, and the unbounded power of explanatory knowledge.
- **Freeman Dyson — Disturbing the Universe:** A memoir of ideas, imagination, and humanity's possibilities.
- **Lewis Dartnell — Origins:** A synthesis of how Earth's history influences human history.
- **NASA & DOE Public Science Resources:** Accessible explanations of fusion, fission, cosmic evolution, and planetary science.

ABOUT THE AUTHOR

Terry M. Adair, DDS, earned his dental degree from the University of California, San Francisco in 1987. He practiced general dentistry in Sacramento, California for twenty-six years before returning to academic training in 2017, where he completed his certification in oral and maxillofacial radiology at the University of Florida in Gainesville, serving as Chief Resident.

Dr. Adair currently resides in Incline Village at Lake Tahoe. Outside of his professional work, he maintains an active, multi-sport lifestyle that includes triathlon training, snowboarding, paddle boarding, mountain biking, open-water swimming, and boating. He is the father of four adult children and shares his home with his cat, Maxwell.

www.ingramcontent.com/pod-product-compliance
Lightning Source LLC
LaVergne TN
LVHW100315210926
848480LV00003B/1124
9798902225324